ORIGINS:

Today's Science, Tomorrow's Myth

ORIGINS:
Today's Science, Tomorrow's Myth

James E. Strickling

PERIPHERAL VISION
Norcross, Georgia

Excerpts from *Earth In Upheaval* by Immanuel Velikovsky © 1955 by Immanuel Velikovsky, reprinted by permission of Doubleday and Co., Inc.

Permission to reproduce excerpts from *The Origin of Races* by Carleton S. Coon has been granted by the Estate of Carleton S. Coon.

Permission to reproduce excerpts from *The Nature of the Stratigraphical Record* has been granted by Macmillan, London and Basingstoke.

Permission to reproduce excerpts from "In the Beginning God Created," by R.F. Surburg, from *Darwin, Evolution, and Creation*, ed. by P.A. Zimmerman, has been granted by Concordia Publishing House, St. Louis, Missouri.

Material from "Who Are the Hairy Ainu?" from *Mysteries of the Past*, ed. by J.J. Thorndike, Jr., is reprinted by permission of American Heritage Publishing Co., Inc., New York, N.Y.

Permission to reproduce excerpts from *Physical Anthropology*, by P.L. Stein and B.M. Rowe, has been granted by The McGraw-Hill Companies, New York.

SECOND EDITION

ISBN 0-9651717-0-1

Library of Congress Catalog Card Number 96-92210

Man will occasionally stumble over the truth, but most of the time he will pick himself up and continue on.

-- Winston Churchill

CONTENTS

ON TURNING STONES

(Foreword to the First Edition)

That all is not well with what can best be termed orthodox science had been hinted at even before the nineteenth century inception of uniformitarianism. The clash between science and religion predates Darwin's theory of evolution. The rift can perhaps be traced to the falling out between William Whiston (1667-1752) and Isaac Newton (1642-1727), both of whom were religious men, and both of whom had attempted to uphold religious truth through the formulation of novel scientific theories.

The cosmic catastrophism that Whiston preached has never died. It but lay dormant for periods of time, rearing its stubborn head now and then, driving true thinkers to newer and alternative hypotheses. Non-catastrophic science has been given no respite. It has had to re-arm itself time and again in its never-ending battles against this multi-headed dragon - and while it has quelled it, and even banished it to the limbo of the scientific fringe, it has not yet vanquished it. Nor will it ever be able to. Try as they may, uniformitarians simply cannot erase the catastrophic signatures with which ancient man signed his oral and written records and with which the Earth itself has imprinted its tortured and twisted strata.

Neither has the cosmos been silent. The astronomy textbooks of yesteryear are now but fiction. What we were once so sure we knew, turned out not to be so. Peering through our telescopes prior to the space age filled our books with nothing but misinformation. It has taken the close planetary fly-bys of man-made hardware to make us see our astronomical errors. And yet, we seem to have learned nothing about our fallibility. We have invented no hardware that could transport us to the past; but while peering at the present, we have the audacity to presume that nothing could have happened then that does not happen now.

I will not invoke the theories of Immanuel Velikovsky that inspired both James Strickling and myself to our present pursuits, for that would involve a debate that is strictly outside the scope of the present work. Personally, I will grant that Velikovsky made mistakes and that his detailed history of cosmic catastrophes has to be drastically revised. But I still maintain that the core of his work is valid -- that the planets of the Solar System were not always at peace with each other and that, in their celestial frays, they committed the Earth to devastating cataclysms.

I will even go further. I claim that the planetary system that our ancient forefathers inadvertently described was an entirely different, and quite bizarre, celestial arrangement. What we see portrayed in the sky today pales to insignificance when compared to what the ancients claim *they* saw. Such ancient portrayals of the cosmos -- descriptions of catastrophic events in the sky as well as down here on Earth -- might easily be relegated to the realm of fantasy were it not for the consistency with which the peoples of the entire Earth recorded them. And it is this that has never been stressed adequately enough. For if these tales were nothing but the fruit of a fertile imagination, how is it that all nations and all races imagined the same events and, quite often, in the same chronological sequence?

James Strickling belongs to the new wave of post-Velikovskian neo-catastrophists. His disclosures on certain subjects have won him a place in the pages of this newly reconstructed science. His insights concerning the nonorigin of language differences (based on the much misunderstood *Genesis* account), as well as his revelations on the effects of Extremely Low Frequency electromagnetic fields (as they apply to the Babel myths), took me completely by surprise when he first brought them to my attention. I am almost envious I did not think of them myself.

But more important is Strickling's call for scientific fairness in the ongoing debate between uniformitarians and Biblical fundamentalists.

Those who know my own writings are well aware that fundamentalism is not my plying trade. Strickling is however correct when he claims that both schools have some merits as well as an awful lot of faults. He is also correct in that the best weapon against Biblical fundamentalists is the Bible itself. But by the same token, science remains the best weapon against scientific uniformitarians. While the truth is sometimes easily recognized, it is often difficult to accept. And in the debate in question, both parties have been stubborn; both have held fast to their dogmas when they should have allowed themselves to be ruled by facts.

I cannot, in all honesty, endorse every statement and hypothesis made and presented in this work. Nor would its author expect me to. In fact, he is well aware of my disagreements on certain issues. It would not be fair to him, or to the reader of this work, to specify what these disagreements amount to. This is not the place for it. Not only has Strickling the right to speak his own mind, but the reader, also, must judge what he reads on the merits of the evidence and arguments presented. What *has* to be stated is that I *do* stand behind the general thesis presented by Strickling and, despite the disagreements mentioned above, I must therefore recommend his work.

Fair enough -- there will be those among the erudite who will find fault with some of the concepts contained in this book. But this much can be said: If stones remain unturned, no sun can ever shine on the ground beneath them.

I will not say that Strickling has been exhaustive in his trek of discovery. He himself will be the last to claim such a thing. The ground he chose is much too vast for any one man to cover. Even so he *has* turned quite a few stones - and uncovered the odd surprise to boot.

My only hope is that those whose expertise enables them to finger the few sore spots that such a work must inevitably contain will not repeat Velikovskian history by throwing out the good with the bad.

Dwardu Cardona
AEON Senior Editor
Vancouver, B.C.
August 22, 1985

PREFACE

You shall know the truth, and the truth will make you free.

Unfortunately, there are many areas in our body of knowledge today where we only *think* we know the truth. And these unrecognized voids make us blind to our ignorance. They stifle us by obviating our need to probe further, since we think "We already know."

"Surely such areas are few in number," you may say. Perhaps. But their impact upon our lives is to be gauged not by their number, but by their pervasiveness within our culture. If a single pothole in the road is large enough, it might throw us into a ditch -- or swallow us up entirely. The "pothole" I address here is our prevailing *worldview*.

The core of the industrial world's twentieth-century worldview is Darwinism -- as in Charles Darwin, who is typically, although erroneously, perceived as "the father of evolutionary theory."

Bear with me. This is not, as you might think at first, a matter for scientists and philosophers only. More likely than not, you're involved to some degree. No matter whether you're being entertained by *Jurassic Park* or being educated by *National Geographic* and *Time-Life* or fighting the school board over the use of a particular high school science textbook, Darwinism is the common denominator. Moreover, whatever code of ethics you choose to follow, it is a subset of your worldview. Our worldview has a direct impact upon how we conduct our daily lives and the values we impart to our children. (Darwinism itself is a subset of a broader philosophy, which this book explores in depth.)

I charge Darwinism, as originally promulgated and currently perceived, with being *The Great Lie*; and it engulfs us. (I am not saying, however, that all of its adherents are guilty of some kind of conspiracy.) Having said that, I must add that I cite Creationist interpretation as being *The Great Mistake*.* So you probably ask at this point, "What's left?" You may be amazed when you find out.

As you continue, don't develop a mental block by telling yourself that you're too ill-equipped to evaluate these matters because you aren't a scientist. Sometimes "seeing the light" comes from merely being "shown the switch." And understanding all electrical theory isn't a requirement for flipping it on. I offer here objective facts and information that have led to my own convictions and insights -- material you can use to make your *own* judgments. After we "flip the switch," I hope your light will gradually become brighter and brighter.

Like so many areas of our society today, "Establishment Science" must remain politically correct, contrary to its avowed objectivity, else there follow dire consequences -- as I have documented in this study. We can expect the reaction of this august body to my presentation to be highly emotional, as it has been to similar works in the recent past. Many careers and intellectual investments are threatened by the introduction of a new paradigm, or worldview. Reasoned debate becomes difficult. When the paradigm shift occurs, from the perspective of this group it will amount to a psychological earthquake of catastrophic proportions.

Even if we can't reach the full truth (hopefully we'll come close), we can winnow many of the prevailing falsehoods.

My subject is origins -- beginnings. In addressing this subject, I appeal not to authority nor to consensus, but to reason. It is dogma in both religion *and* science that appears to be at the very heart of the ongoing controversy surrounding our origins. Reason, I believe, can be the antidote to dogma and emotion. It can illuminate, and possibly fill, those heretofore unknown voids in our knowledge.

---

* As far as intellectual understanding and advancement are concerned, I submit that the Creation-Evolution controversy in the classroom is a lose-lose situation.

INTRODUCTION

The Subject

Our solar system consists of nine major planets plus numerous moons, asteroids, comets and dust, all orbiting the sun in a sea of electromagnetism. The members show wide variations in size, composition, spin rate, inclination of axis, and direction of movement. There are few aspects that could be called uniform, and a tumultuous history is indicated.

Planet Earth is unique within this assemblage. A beautiful blue and white gem as seen from afar, it is covered with oceans of water and teems with life; one cannot help but marvel at its variety and complexity of relationships. The Apollo 8 astronauts appropriately described our world as an oasis in space.

Once believed by Western man to be at the center of the universe, the earth actually appears to be a lonely beacon of intelligence at the outskirts of the Milky Way, far, far from the center of anything. It is not even at the center of the solar system.

Mankind's apparent aloneness in illimitable expanses provokes the age-old queries: Where did everything come from?

Was there ever a beginning? Or is existence part of an eternal cycle? And life: Is it an inevitable product of matter and physical law?

Or did it come from some kind of Divine Spark?

All cultures have asked these questions, and most have arrived at answers that satisfied them, faithfully preserved, in times past, in myth, legend or holy writ. Modern science, however, has challenged the ancient authorities; recourse to the supernatural, for most contemporaries, is a thing of the past. But does the scientist really have the answers? Will Truth rest ultimately with him or with the theologian? Can the truth even be known?

We continue to ask. We continue to search. Today the search is manifested in a great debate that is molded to its particular shape through the fact that Man is an intrinsically subjective being; it is difficult for human beings to perceive objectively. Everything we experience is somehow affected in our minds by previous experience; no area of our lives is excepted. Subjectivity is carried into all of our institutions, both small and great.

Religious institutions, most people would agree, are highly subjective. In Christianity, this fact is driven home by the multiplicity of church denominations: They all grew out of different interpretations of the same book.

Less obvious to the general public is that "science" suffers from the same impediment; many scientific theorizers, though attempting to be objective, have relegated objectivity to the List of Endangered Species. The creation-evolution controversy -- the great metaphysical debate of our time -- provides a case in point. Theory, internalized as a mode of seeing and believing, stands between the scientist and many of the facts adduced against traditional evolutionism.

The situation is not new: sharp polarization, with many instances of tunnel vision on both sides of the controversy.

The issues did not begin with Charles Darwin; they are far older, and the debate is far from over, for neither side can draw the least concession from the other. Evolutionists and creationists both claim to have proven their case. Yet they are totally at odds with one another. Who is correct? Both? Neither?

Creationist viewpoints today are subsumed under the heading "Scientific Creationism." Included are several other concepts in addition to the acts and methods of creation. Whatever the implications of the word "scientific," all disputants understand that the term incorporates and re-interprets "biblical creationism." Not apparent to either its adherents or its major detractors is that scientific creationism is *not* biblical -- not, at any rate, in its entirety.

The evolutionist position, as publicized and as it stands in the mind of the general public, rests on Darwinism. Darwin's only distinctive contribution, however, other than his careful marshaling of data around a central thesis, was the idea that "natural selection" is the mechanism of evolution, and many scientists have begun to realize that such a mechanism is inadequate to account for the observed facts. The concept is being propped up by continual speculation and *ad hoc* interpretations *ad infinitum*. Yet because Darwin cannot be allowed to fall from official grace, the sterility of the concept of natural selection in explaining the production of new species remains concealed from popular view.

Neither side of the controversy appears capable of recognizing its own weaknesses, or that the two factions share some of the same weaknesses. Darwinism declares that "natural selection" among random changes, spread over many eons, accounts for all of the life- forms on Earth. Fundamentalism declares that an act of God produced each species, just as acts of God produced the laws of nature. The Darwinian position, like the Fundamentalist one, is built on faith: No Darwinian has ever seen the evolution of a life-form, any more than any Fundamentalist has ever seen a life-form created.

The Christian avowedly rests his or her position on faith. The Darwinian talks a different game -- probabilities. The Darwinian relies on probabilities without having ever established a series of events on which to base a probability. To the Darwinian one might say: "You have not even established the existence of a working law. The applicability of an axiom, such as Newton's axiom about the force of gravity, is ascertained only after many observations and measurements. The axiom is fitted to the observable behavior of nature. The applicability of a theory is ascertained in the same way. But you have not seen your theory work even once. How can you argue the probabilities of successful mutation to produce many new species when you have not seen mutation produce even *one* new species?"[1]

Given the obvious weaknesses of Darwinism and Creationism, it is time for an unbiased examination of both. If the weaknesses prove crippling to the hypotheses of both factions, we should be willing to look for an alternative hypothesis.

There is, I believe, an alternative, but since I am its proponent, I must ask my readers to suspend judgment temporarily and maintain the disinterested view that necessarily was surrendered by the author in forming rival hypotheses to those on both sides of the present debate. (Like the Darwinian, like the Creationist, I was not around to see the events that I postulate. Like their theories, therefore, mine is a reconstruction of the distant past.)

As a first step toward weighing all the hypotheses, my readers will have to go behind Darwinism and Creationism to a less familiar controversy: Uniformitarianism vs. Catastrophism. And they will have to be prepared to weigh a fundamental contention of this book: that within the definitions of these jaw-breaking terms lives an issue that can provide the foundation for an intellectually satisfying alternative theory, neither creationist nor Darwinian, that is logically sound and free of self-contradictions. In turn, this alternative can lead to new paths of investigation into the worlds around us.

The investigation that follows will embrace a great body of evidence scarcely ever considered -- and then only very selectively -- by either creationists or evolutionist-uniformitarians. Recognized by the combatants only insofar as it lends strength to conclusions already reached is the world's enormous store of mythology. Is all myth to be regarded as fabrication? If not fabrication, is myth perhaps fanciful exaggeration of real facts? Or might it instead recall real experiences for which there were hardly any words for expression?

A *very* great many myths speak of things unfamiliar to historical man. Should they be filtered and interpreted based on contemporary scientific theories and experience? Or would that be putting the cart before the horse?

Until the eighteenth century, theory and "collective experience" ruled that stones could not fall from the sky. Meteors continued to fall, however, and theories finally had to be reformulated.

How much information is being denied us today by our theories?

When John Milton wrote *Paradise Lost*, he could not decide between the Ptolemaic and post-Copernican versions of the universe, so he hung the Ptolemaic (or "created") system inside the larger (or "aboriginal") system. In other words, he waived the question of origin of the universe of stars and concentrated only on mankind's immediate world, the solar system. The rest was "Chaos and Old Night."[2]

This poetic solution involved a willingness to live with some uncertainty while dealing with realms of experience that seemed, at least, to be malleable to the shaping powers of logic, analogical inference, observation, imagination, and reason. It was akin to accepting that a bumblebee can fly even though, aerodynamically, it should not be able to do so. Or to put the matter differently: It was an acceptance of limits, a relaxation of intellectual arrogance. It was a concession by a learned man that there might be more things in heaven and earth than were dreamed of in his philosophy.

Time is as difficult to penetrate and know as is space. If Milton could concede that space might contain much more darkness than his lamp could illumine, perhaps a modern reader predisposed to accept Darwinism uncritically might be willing to make the same concession about time.

This study is an excursion into the distant past. It is a search for beginnings and for what came before -- for times and phenomena that exist now only as faint memories, if at all. It is a hunt for ancient terrestrial and celestial upheaval, for flood and destruction, for creation from chaos. It is a search for alternatives to an outmoded worldview.

Our planet appears to have had a far stranger history -- and a more precarious one -- than science is willing to recognize. Indeed, from within the constraints of uniformitarianism, this recognition may be hardly possible.

Uniformitarianism asserts that nothing has happened in the earth's past that is not happening now. All geological and biological change takes place, and has always taken place, at an imperceptible rate. Only over periods of millions of years can a difference be noticed -- evolution from microbe to man, the growth of a mountain range. The gradualist looks at the world around him and makes assumptions based on what he sees. He uses these assumptions to bind his mind.

A purpose of this book is to expose the fallacies of such a method. Gradualism, I believe, cannot account for the geological record. Underlying the arguments that follow is the premise of *global catastrophism*, the idea that in times past the earth has suffered numerous disruptions from the nearby presence of other planet-sized bodies. Such disruptions would impact not only the earth itself, but all the life-forms residing on it. Evidence of these disturbances abounds.

Many of the chapters in this work draw heavily from biblical records; some chapters find their very basis in accounts that many people mistakenly believe to occur only in the Bible. In fact, the Bible is treated here exactly as any other ancient literary source. It is not employed as an infallible standard by which other information is accepted or rejected, nor is it treated lightly as religious imagery, pseudo-history or fabrication. Rather, it is one more voice to be heard.

The subject to be explored is origins -- beginnings -- of many different aspects of our experience. Creationists and Darwinians draw upon their respective theologies (Christian and secular) for ready-made "answers" to all questions about beginnings. The appeal of this work is to reason, not theology.

The hypotheses advanced in this book may sound strange by comparison with popular "scientific" notions -- but if so, they sound strange mainly, I contend, because they are unfamiliar concepts. However odd-sounding and whatever their weaknesses, these alternatives to conventional views are derived through an honest examination of both physical (i.e., geological) and cultural records, which, taken together, reveal more than either can do alone and independently. The ideas advanced in the forthcoming pages stand opposed to many of the notions perpetuated in our educational institutions. As you read, I ask only for an open mind, a willingness to weigh all of the evidence. I invite you to judge for yourself.

REFERENCES

1. The argument in quotations is adapted from a similar one in a passage of Platonic dialogue by Morris Kline in *Mathematics in Western Culture*, Oxford University Press (London 1953), pp. 388-9.

2. See, for example, *The Norton Anthology of English Literature*, Vol. 1, W. W. Norton & Co. (New York 1968), p. 1039, note 8, explaining Milton's two images of the cosmos.

PART ONE

Creationism, Evolution, and Catastrophism

CHAPTER I

The Uniformitarianism-Catastrophism Controversy

As Charles Darwin contemplated the diversity of life-forms that have come and gone on this planet, he was driven to reflection on a global cataclysm; this, he felt, would explain the several great worldwide extinctions of life attested by the fossil record. He was, as he explained, "irresistibly hurried into the belief of some great catastrophe; but thus to destroy animals, both large and small, in Southern Patagonia, in Brazil, on the Cordillera of Peru, in North America up to the Bering's Straits, we must shake the entire framework of the globe."[1]

Instead of accepting the implications of the observed facts, however, he concluded that his observations were not indicative of what actually had happened. "Thus[,] he theorized that the geological and paleontological record was really incomplete and compressed and abbreviated, so that what is continuous only *appears* discrete, and what is really slow only *appears* fast, and what is really non-simultaneous only *appears* simultaneous."[2] In all of these respects, Darwin reasoned contrary to the known evidence.

Darwin endeavored to show that what appeared to be the result of global upheavals was to be explained as the product of slow changes through eons of time, with no intervening violence, urging that "a man should examine for himself the great piles of superimposed strata, and watch the rivulets bringing down mud, and the waves wearing away the seacliffs, in order to comprehend something about the duration of past time. . . . Nothing impresses the mind with the vast duration of time, according to our ideas of time, more forcibly than the conviction thus gained that subaerial agencies which apparently have so little power, and which seem to work so slowly, have produced great results."[3] .

Why did Darwin go against his own field observations? A bit of intellectual history is in order. Charles Darwin was not really the beginning of something; he was an important *link* in a chain of speculation about origins -- indeed, his grandfather, Erasmus Darwin, had been a much-ridiculed link.

Christianity itself, assimilating Aristotle, Plato and Plotinus into its thought, prepared the way for Darwinism. Christian theology came to see nature as an infinite series of forms, from inanimate to animate, from the simplest life-form to man, and from man (placed at a crucial point in the chain of being by virtue of his reason and spiritual gifts) to the angels and other spirits, and from these to God.

The leading principles of this theology were plenitude and continuity.[4] These principles, seen as expressions of the goodness of God, had definite implications for the study of species. In *De Animalibus*, Albertus Magnus (c.1200-1280) said that "nature does not make [animal] kinds separate without making something intermediate between them . . ."[5] In his turn, Thomas Aquinas (1225-1274) marveled at the "wonderful linkage of beings" that nature "reveals to our view. The lowest member of the higher genus is always found to border upon the highest member of the lower genus."[6]

Another expression of this worldview was given by the eighteenth century English poet Edward Young. Man, Young wrote, is a "Distinguished link in being's endless chain, Midway from nothing to the deity."[7]

The use that Young made of the chain of being was somewhat atypical; he used it to permit man to hold a rather high opinion of himself. Most theologians, poets and divines used it to inspire humility in Christians, since, as Joseph Addison noted, the "space and room" upwards from man was infinite, while the number of grades below was finite.[8] As Arthur O. Lovejoy expresses Addison's point, man "was the 'middle link' [in the chain of being] in the sense that he was at the point of transition from the merely sentient to the intellectual forms of being."[9] Thus, man was well toward the lower end of the scale of being and had no reason to flatter himself.

However, there were inherent logical weaknesses in the Great Chain of Being, and these began to be exposed at the same time that seventeenth and eighteenth century science was asserting its sway over European and British thought. Suddenly, *philosophes* and other freethinkers began to turn the Great Chain of Being against itself -- and to turn it upside down. Instead of the chain of being *descending* from God down to nothingness, it was made to *ascend* from the veriest speck in the void up to man, where, abruptly, it stopped.

As early as 1749, in his *Histoire naturelle, general et particuliere*, the Frenchman Georges Louis Leclerc de Buffon (1707-1788), working from the principles of "universal continuity and necessity" (in the old orthodoxy, the principle of plenitude had been seen as a type of "necessity"), argued that (in the words of Will and Ariel Durant) "there are no fixed and unchangeable species in nature, that one species melts into the next, and that science, if matured, could ascend step by step from supposedly lifeless minerals to man himself."[10] And he added:

> [I]f we once admit that there are families of plants and animals, so that the ass may be of the family of the horse, and that one might differ from another only by degeneration from a common ancestor, . . . we might be driven to admit that the ape is of the family of man, that he is but a degenerate man, and that he and man have had a common ancestor. . .[11]

He argued that new varieties of animals might be produced by geographical migration and segregation and that many species of animals had been destroyed, perfected or made degenerate by "great changes in land or sea, by the favors or disfavors of Nature, by food, by the prolonged influences of climate, contrary or favorable. . . ."[12] From Buffon's theory to that of Charles Darwin is but a small step.

Darwin's theory was foreshadowed by that of Pierre Louis Moreau de Maupertuis (1698-1759), who in 1751 wrote:

> The elementary particles which form the embryo are each drawn from the corresponding structure of the parent, and conserve a sort of recollection of their previous form. . . .

> We can thus readily explain how new species are formed, . . . by supposing that the elementary particles may not always retain the order which they present in the parents, but may fortuitously produce differences which, multiplying and accumulating, have resulted in the infinite variety of species which we see at the present time.[13]

Maupertuis conjectured that, in the fullness of time, a single prototype could have produced all living forms. Thus, a hundred and eight years before Charles Darwin published his *Origin of Species*, Maupertuis postulated evolution from cumulative mutations. In the same year (1751), the English philosopher David Hume speculated in his *Dialogues concerning Natural Religion* that "the adaptations of organs to purposes may have resulted not from divine guidance but from nature's slow and bungling experiments. . . ."[14] "Slow and bungling experiments" are tantamount to Darwin's theory of natural selection.

In the century from Maupertuis and Hume to Darwin, theories of evolution became more fully elaborated. Along the way, however, there appeared a few surprising developments. One of these was the evolutionary theory of Erasmus Darwin (1731-1802), grandfather of Charles. Erasmus postulated an expansive "nisus" in all forms of life that leads each form to strive for an ever higher expression.[15] (The word "nisus" is Latin for "a pressing, straining, or effort."[16]) Erasmus' mystical explanation of the evolution of all life-forms from a "primal filament" encountered much ridicule, including that of an influential conservative publication, *The Anti-Jacobin*, which in "The Loves of the Triangles" in 1798 heaped scorn on the grandfather of the author of *Origin of Species*. Wrote the *Anti-Jacobin* authors in one of their mock "notes" to "The Loves of the Triangles":

> The ancient mathematicians conceived that a cone was generated by the revolution of a triangle; but this, as our author [the fictive Mr. Higgins] justly observes, would be impossible, without supposing in the triangle that expansive nisus, discovered by Blumenbach, and improved by [Erasmus] Darwin, which is peculiar to animated matter, and which alone explains the whole mystery of organisation.[17]

Because of Erasmus Darwin's unfortunate foray into evolutionary theory (and until publication of *Origin of Species*), the word "Darwinian" was used as a term of ridicule, and the poet Samuel Taylor Coleridge coined the term "darwinizing" to describe wild evolutionary theorizing.[18]

When Charles Darwin came along, he would naturally incline toward a theory that would not encounter the scorn with which his grandfather's notions had been met. Purposive evolutionary striving by microorganisms, plants, and lower forms of animals, vis-à-vis the Chevalier de Lamarck's theories or Erasmus' unfortunate musings, would never do. Nor would Creationism do. In *The Autobiography of Charles Darwin*, he made his feelings explicit: "I for one must be content to remain an Agnostic."[19] Of mankind's tendency to believe in a First Cause (or a god), he wrote that it "probably depends merely on inherited experience" and added: "Nor must we overlook the probability of the constant inculcation in a belief in God on the minds of children producing so strong and perhaps an inherited effect on their brains not yet fully developed, that it would be as difficult for them to throw off their belief in God, as for a monkey to throw off its instinctive fear and hatred of a snake."[20]

This left Darwin a choice of two rival theories: cataclysmic destruction and creation of species; or the gradualism of natural selection.

As it happened, cataclysmic theories had met with ridicule also, as in the case of Maupertuis, who (in an attempt to fill up the gaps in the chain of being) suggested that many species of life once existing must have been eliminated by some celestial accident, such as the approach of a comet.[21] It was primarily the religious-minded who toyed with ideas of catastrophic collision with comets or asteroids, as S. Austin Jr. did in "The Comet" and Edgar Allan Poe did in "The Conversation of Eiros and Charmion"[22] (These stories were concerned with the biblical prophecy of the end of the world by fire.)

By contrast, for a naturalist like Darwin the way had been well-paved -- by Maupertuis (in his evolutionary thought), Buffon, Lord Monboddo, James Hutton, Denis Diderot and others -- for the principle of gradualism and the theory of random mutation.[23] Indeed, evolutionary thought had penetrated popular culture. Eight years before *Origin of Species* was published, the poet Alfred Tennyson, in his elegy *In Memorium*, had written:

". . . From scarped cliff and quarried stone" She [nature] cries,
"A thousand types are gone; I care for nothing, all shall go . . .
. . . I bring to life, I bring to death . . ."[24]

Even earlier, on the other side of the Atlantic Ocean, P.T. Barnum, one of the most remarkable hoaxers and scam artists in American history, was luring people to his American Museum to see "missing links." In 1842, the master of chicanery advertised "the preserved body of a Feejee Mermaid" (supposed to be a connecting link between man and fish) and such specimens as "the Ornithorhincus, or the connecting link between the seal and the duck; two distinct species of flying fish, which undoubtedly connect the bird and the fish; the Siren, or Mud Iguana, a connecting link between the reptiles and fish, . . . with other animals forming connecting links in the great chain of Animated Nature."[25] The so-called Feejee Mermaid actually was the top part of a mummified monkey attached to the bottom part of a dried fish.[26] The point is not that Barnum was a master of humbug, but that he was a wonderful judge of what the public wanted to see. In 1842, the public already had a huge appetite for the artifacts of evolutionism -- even fake ones.

Thus, Charles Darwin did not publish *Origin of Species* in an atmosphere that was unprepared for his theory. In fact, so thoroughly was the way prepared that in 1858, Darwin and Alfred Russell Wallace, having formulated their theories independently of one another, published separate papers in the same issue of *Journal of the Proceedings of the Linnean Society* in which they expressed exactly the same theory of natural selection.[27] The next year Darwin published his most famous book, and the entire first edition (1,250 copies) sold out on the first day.[28]

The background of Charles Darwin's work on *Origin of Species* is important in understanding how readily Darwin would turn to James Hutton's principle of uniformitarianism, which was simply a time-tinctured elaboration from the principle of continuity. And the notion of continuity, of course, had been widely accepted in Western man's conception of the "ladder" or "chain" of life even before the eighteenth century skeptics, naturalists and philosophers turned the Chain of Being upside down. The background thus also explains why so many people were prepared to accept the Darwin/Wallace theory of random selection in the evolution of species.

In brief, the rapid dissemination of Darwinism grew out of two centuries of speculation about the Great Chain of Being. To that extent, it was a historical accident.

The Same Now and Always

Uniformitarianism proclaims that "the present is the key to the past." It asserts that nothing of geological or evolutionary significance has happened in the past that is not happening right now.* William Stansfield wrote that "the forces now operating in the world are those that have always operated, and . . . the universe is the result of their continuous operation."[29]

It has been observed that Darwin's theory was declared to be "progress as compared with the teachings of the Church. The Church assumed a world without change in nature since the Beginning. Darwin introduced the principle of slow but steady change in one direction, from one age to another, from one aeon to another. In comparison to the Church's teaching of immutability [of species], Darwin's theory of slow evolution through natural selection or the survival of the fittest was an advance . . ."[30]

The average church congregation, in an age when most people attended church, accepted the biblical account of Creation. nineteenth century European and British culture was space- and time-haunted and was kept in turmoil by the ideas of atheists, skeptics, utilitarians, social and political revolutionaries, biological and geological scientists, and other free-thinkers, but the society in turmoil was dominated numerically by conventional believers in Genesis.[31] Known fossils were attributed by most people to Noah's flood. Darwin's writings, especially *Origin of Species* and *Descent of Man*, helped spread the gospel of evolution, and the gospel of Genesis suffered. In the schools, the Bible was displaced and Darwinism entrenched by the 1930s. Today, uniformitarianism is seldom seriously questioned in our educational institutions; in textbooks, it is presented as absolute fact.

---

* The full scope of uniformitarianism is now often divided between two terms, "uniformitarianism" and "actualism," distinguishing between types of processes and rates of process. Throughout this discussion, "uniformitarianism" is used to express both ideas, as this term is the more familiar of the two.

But educational texts, even in science and mathematics, tend to lag behind the times. As Morris Kline wrote in 1953, "The calculus textbook most widely used in the United States during the last fifty years and the one that is still most popular might well have been written in 1700," despite the fact that since Newton's day calculus had been perfected and had become the cornerstone for *analysis*, "a branch of mathematics, far vaster than algebra and geometry, that has served, guided, and led science so remarkably."[32] A similar situation exists today in science books that blandly present classical Darwinism without noticing that the conventional evolutionist views are being modified.

An example of these modifications has to do with catastrophism. Reputable scientists no longer are scandalized by suggestions akin to that of Maupertuis (that a "comet" striking the earth could have extinguished many life-forms). Nobel Laureate Luis Alvarez and his son Walter argued that the earth probably was struck at the end of the Cretaceous Period by a Manhattan Island-sized asteroid and that this collision may have had a great deal to do with the extinction of dinosaurs.[33] While other researchers feel the asteroid impact most likely was only a contributing factor to the "great extinction," so esteemed a scientist as Stephen Jay Gould of Harvard University believes that "the astronomical catastrophe may have been more than just a minor *coup de grace*. After all, extinctions far less extensive than the Cretaceous event have punctuated the history of life many times during the past 600 million years. Perhaps they never really have great impact unless the general deterioration of conditions that serves as their major cause is *greatly amplified* in effect by some *extraterrestrial* event."[34]

With Darwinian orthodoxy beginning to relent a little, now may be the time to hope for a serious reconsideration of some basic tenets of uniformitarianism. However, the Darwinian viewpoint is rooted in a faith, not a disinterested science, and catastrophist heresies continue to receive a hostile reception. Of Darwin and his followers, Lynn E. Rose has written: "A safe uniformitarian Earth was, for them, much preferable to the repeatedly devastated world that is indicated by the geological and paleontological facts."[35] Darwin himself wrote that "we may feel certain . . . that no cataclysm has decimated the whole world. Hence we may look with some confidence to a secure future of great length."[36] Reading such hopeful words, one might easily be reminded of the eighteenth century mathematician Jean le Rond d'Alembert (1717-1783), who, reassuring his students about the unreliable state of calculus, advised them to continue their study of it; *faith*, he said, would come to them.[37]

The Catastrophist Alternative

For now, the gauge of the scientific community's receptiveness to catastrophism remains the career of Immanuel Velikovsky. Outside the camps of Velikovsky's dedicated supporters and detractors, relatively few people recognize the name of this man. Fewer still are aware of the implications of his work. But several decades ago, Velikovsky stirred up a scientific ruckus, that continues to this day, when he dared to question the doctrine of Uniformitarianism by propounding a theory that revived the concept of global catastrophe -- not only in the study of the geological record, but in early historical records as well.

Velikovsky's thoroughly researched *Worlds in Collision*, published in 1950, was the first of several of his books that cut across the fields of physics, astronomy, geology, archaeology, paleontology, history and mythology, shaking the foundations, some people felt, of every discipline touched. When *Worlds in Collision* appeared, many leading members of the scientific community (Albert Einstein being a notable exception) branded Velikovsky a crackpot. A few scientists weighed his claims objectively, but many rejected Velikovsky's ideas without fully understanding them -- in some cases, without even knowing what they were. Velikovsky was ridiculed, libeled and finally, to the extent possible, ignored. Scientific investigations have substantiated a number of his hypotheses and predictions, but official recognition is still being withheld in the leading journals of science.[38]

In brief summary, what Velikovsky said was this: He asserted that the solar system, as well as the rest of the universe, was electrical in nature (who would question this today?). Electrical effects, he said, explain many phenomena not otherwise explicable. He also alleged that the structure of the solar system has changed in historical times. Some of the other planets once were much closer to our own and were envisaged by ancient man as frightening gods or monstrous animals. The changes leading to the present order had cataclysmic effects upon the earth; even as Darwin had mused, the entire globe *was* shaken on its axis -- more than once.[39]

These were not idle speculations on Velikovsky's part; he accumulated massive amounts of scientific and historical evidence to support his hypotheses. And as a result of his research and analysis, he was able to make many predictions that have been borne out.

One of these was his prediction of radio noise emanating from Jupiter. Einstein said that fulfillment of predictions such as this would go far in establishing the credibility of Velikovsky's theories.[40] Six months after Velikovsky made the prediction, the signals were detected. Unfortunately, Einstein died before he could marshal any support for Velikovsky or call the world's attention to the signals having been anticipated. When he passed away, a copy of *Worlds in Collision* lay open on his desk. When the discovery of radio noise from Jupiter was announced, it was claimed to have been unexpected.[41]

Velikovsky's successful predictions are too many in number to detail here, but they span planetary electromagnetism, comets and meteorites, Mercury, Venus, Earth, the moon, Jupiter, Saturn, Pluto, and ancient history and prehistory.[42] In *American Behavioral Scientist*, it was stated of Velikovsky's work: "Seldom in the history of science have so many diverse anticipations -- the natural fallout from a single central idea -- been so quickly substantiated by investigations."[43]

Never mind; Velikovsky was savaged. Professional scientists launched a campaign against *Worlds in Collision* before it was even published. The attack was led by Harlow Shapley, who arranged for denunciations of the book by scientists who had never read it. Shapley and others forced the dismissal of James Putnam, the trade-books editor of the Macmillan Company responsible for accepting the Velikovsky manuscript, and of Gordon Atwater, curator of New York's Hayden Planetarium. Atwater's crime was in taking Velikovsky seriously enough to program a planetarium show based on his theories. After Shapley threatened Macmillan's textbook contracts, Macmillan finally prevailed on Velikovsky to let it transfer its rights in *Worlds in Collision* to a competitor, Doubleday, which had no textbook division and thus was not susceptible to professorial blackmail.[44]

The behavior of the scientific establishment in Velikovsky's case is exactly analogous to that of the same establishment when Galileo (1564-1642) announced his belief that the behavior of sunspots proved that the earth revolves around the sun. The Aristotelian professors, perceiving their vested interests in the Ptolemaic system to be threatened, united against Galileo and conspired to turn the Vatican and the Inquisition against him.[45]

The scientific establishment, of course, both four hundred years ago and today, has an unlimited capacity to deny its own inquisitorial character. According to a recent biology textbook:

> . . . We cannot imagine that the cause of truth is served by keeping unpopular or minority ideas under wraps. Today's students are much less inclined than those of former generations to unquestioningly accept the pronouncements of "authority." Specious arguments can only be exposed by examining them. Nothing is so unscientific as the inquisition mentality that served, as it thought, the truth, by seeking to suppress or conceal dissent rather than grappling with it.[46]

The furor in the scientific community over *Worlds in Collision* arose in part because Velikovsky was himself a scientist of repute. If he had been aligned somehow with creationists or orthodox theologians, there probably would have been no uproar because he could have been ignored with ease. Exactly the same point can be made about the defenders of the Ptolemaic system in Galileo's time: Galileo was dangerous to the Aristotelian professors because, as a brilliant mathematician and astronomer, he was persuasive.

What makes men like Galileo and Velikovsky dangerous? Their work imperils the entire worldview of other men; their hypotheses are profoundly disturbing. Galileo tore the earth from the center of a finite universe; Velikovsky tried to tear the earth from its comfortable revolutions within a solar system that was predictable over billions of years. Thus, Doctor Uniformitarian asserts that catastrophism on the scale proposed by Velikovsky cannot be supported. Contemporary textbooks, supplanting the papal bull, declare catastrophism untenable.[47]

In fairness to the establishment, perhaps a satisfying orientation -- in science, religion or just plain, everyday living -- is too hard-won to be surrendered lightly. New ideas always have to battle against the inertia of Establishment conservatism. Stephen Talbott has observed that such conservatism has the virtue of enabling people to extract the last breath of "explanation potential" from each accepted theory before it is set aside. A dominant theory thus remains on its throne for as long as it can adapt, even with much straining, to the available evidence. Thus, apart from rare upheavals, the overall progress of science appears to be smooth and steady, with very few truly disruptive false starts or wrong turns.[48]

The disadvantage of such conservatism is that, as the investment of emotion, reputation and research funds increases, mistakes tend to become larger, and a significant change of direction requires a full-scale revolution. Change can become next to impossible.

Talbott goes on to note the different perspectives assumed by contrasting philosophical positions. A scientist who believes the face of the earth to have been molded substantially by catastrophic events and a scientist who accepts the principle of uniformity will regard a body of data differently; each scientist will perceive the data as fitting his or her own interpretive schemes. In fact, the assumptions of each individual may actually determine beforehand the nature and relevancy of the available data and evidence. For example, a geologist inquiring about the age of a particular layer of rock, one assumed by his geological scheme to be a billion years old, might turn to a physicist for "independent" verification. The physicist then would apply radiometric dating methods which themselves were developed and checked against the geologic column using uniformitarian assumptions. To no one's surprise, the radiometric dating would yield "appropriate" dates.[49]

Moreover, the more entrenched and inclusive an assumption or theory becomes, the easier it is to ignore even those facts for which there is absolutely no accounting. Flukes cannot be allowed to overturn the well-grounded theories and/or methods of generations of scientists.[50] Such stubborn grappling can bear marvelous fruit, as when several generations of mathematician-scientists managed to refine and finally perfect the rudimentary calculus of Newton and Leibniz by checking results against physical arguments and actual measurements. It might be useful in casting some light on the current situation to note that because of "the lack of rigor in the early works on the calculus," controversy over the soundness of the new mathematics was intense and prolonged, and mathematician Michel Rolle, a contemporary of Newton, "taught that the calculus was a collection of ingenious fallacies."[51]

The best wisdom for those wishing to challenge uniformitarianism might be to recognize that catastrophism is in the stage of a "lack of rigor" that characterized the calculus for a hundred and forty years before the French mathematician Augustin-Louis Cauchy published a definitive work, *Cours d'Analyse* (1821).[52] In science, a radically new theory generally cannot make inroads until an older, established one begins to fall apart of its own accord, as the Great Chain of Being did at the end of the eighteenth century. There are a few signs, however, that the uniformitarian establishment may be crawling out on a limb that soon will break.

Gould and Richard Lewontin, quoted by Alexander Rosenburg in a philosophical treatise on biological science, give convincing support to Talbott's claim that conventional science ignores evidence not in accord with its theses. Gould and Lewontin provide a revealing example of the uniformitarian mind-set. In their discussion of biological adaptation (that is, the evolution of organisms to become compatible with their environments) the following rules are given as Darwinist methodology:

(1) "If one adaptive argument fails, try another." Antlers are first viewed as adaptations for protection from predators, but when this argument is overturned, they are viewed as adaptations for intraspecies dominance.

(2) "If one adaptive argument fails, assume that another, as yet undiscovered one, exists." The failure to substantiate an adaptive claim is never viewed as a fault in the basic theory but, rather, is seen as a motive to search farther and wider for an argument that will hold up, at least provisionally.

(3) "In the absence of a good adaptive argument in the first place, attribute failure to the imperfect understanding of where an organism lives and what it does." After all, none of us has ever lived under a rotting log, and to move the log is to disturb the habitat of the insects living under it. We must be patient and cunning, and, sooner or later, enlightenment will be ours.

(4) "Emphasize immediate utility and exclude other attributes of form." In particular, adaptationalists appear to ignore opposed explanations even when these seem to be more interesting and fruitful than the preferred untestable speculations.

Gould and Lewontin conclude by stating that they would not strongly object to such reasoning if *in principle* it could lead to a rejection of adaptational theory for a lack of evidence. "[I]f it could be dismissed after failing some explicit test, then alternatives would get their chance. Unfortunately, a common procedure among evolutionists does not allow such definable rejection for two reasons. First, the rejection of one adaptive story usually leads to its replacement by another, rather than to a suspicion that a different kind of explanation might be required. Since the range of adaptive stories is as wide as our minds are fertile, new stories can always be postulated. . . . Secondly, the criteria for acceptance of a story are so loose that many pass without proper confirmation. *Often evolutionists use consistency with natural selection as the sole criterion and consider their work done when they concoct a plausible story"* (Gould and Lewontin, 1979:586-8).[53]

The subjectivity noted by Gould and Lewontin is reflected everywhere in uniformitarian writings. William D. Thornbury claims that without "the principle of uniformitarianism there could hardly be a science of geology that was more than pure description."[54] *Encyclopaedia Britannica*, an authority for countless young students, states that the principle of uniformitarianism has actually been proven in geology,[55] and Windsor Chorlton makes reference to "The Law of Uniformitarianism."[56] Whatever has happened to the bruited objectivity of science? Uniformitarianism is an *assumption* about the past that serves as a basis for interpreting data. It is not provable. It is not a law.

As to its application, this assumption apparently amounts to a blank check. Writes Diane P. Gifford: "All analyses of prehistoric materials involve some kind of assumption of uniformitarianism. . . . Moreover, paleontologists regularly extend the assumption of uniformity well beyond what can be inferred directly from the observable attributes of a preserved element, inferring associations with a specifiable complex of other elements of which no traces actually exist. . . ."[57]

Thornbury relates that when J. Harlan Bretz suggested around 1923 that the channeled scabland area of eastern Washington had been cut out by floodwaters of enormous proportions, his idea was met with much opposition -- only because *"it represented a return to catastrophism."*[58]

Uniformitarianism is not to be violated; it is not debated. Nor is it restricted to terrestrial science; Dwardu Cardona comments that "modern astronomers, who should know better, also look up at the sky, and what they do not see there, they believe could never have existed. What doesn't happen now, they have the audacity to preach, could not have happened then."[59]

The uniformitarian evaluation of catastrophism can be likened to a builder measuring materials with a ruler one foot long divided into 120 equal "inches." Everything measured will appear to be ten times longer than it actually is -- i.e., every catastrophist conclusion will appear to be what people used to call "a stretcher" (a protean lie). The ruler is therefore misleading; the uniformitarian ruler cannot be trusted to evaluate an approach whose basic challenge is to the ruler.

There is a legitimate category of evidence that the uniformitarians, by habit, either minimize or ignore: the testimony of our ancient forebears. In history, in myth, in scripture -- or in whatever vehicle -- the ancient record deserves to be considered, for those who left this testimony behind claim to have experienced the traumatic upheavals that uniformitarians say could not have happened. The echoes of their experience are too widespread to be discounted. And if sweeping catastrophe did overwhelm their world, it is not unreasonable to suppose that global upheavals occurred before their time as well. As for the earlier epochs, the earth speaks for itself; it bears deep signs of its tortured past.

But it does not speak thus, of course, to uniformitarians. Writes Bennison Gray, "no evidence against uniformitarianism [can] be used, because it fails [in the minds of uniformitarians] to qualify as scientific. History is *not allowed* to refute scientific uniformitarianism."[60] Since uniformitarianism is assumed to be true, evidence challenging it must be denigrated to the level of uninformed prattle. Uniformitarianism is a principle.[*] Even the most obvious evidence of global catastrophe is forced to conform to a uniformitarian framework.

But what ensues without such a presumption? And what are the ramifications for the philosophies of Creationism and Darwinism?

---

* "Principle", according to Webster, means a comprehensive and fundamental law (as uniformitarianism is considered to be) or a doctrine/dogma (as uniformitarianism truly is) or assumption (which it is also).

REFERENCES

1. Charles Darwin's *Journal* of the Beagle's Voyage (January 1834).
2. Lynn E. Rose, "On Velikovsky and Darwin," *KRONOS* VII-4 (Summer 1982), p. 38.
3. Charles Darwin, *Origin of Species*, Chapter X.
4. Arthur O. Lovejoy, *The Great Chain of Being*, Harvard University Press (Cambridge 1936, 1964), passim.
5. Ibid., p. 79.
6. Ibid.
7. Edward Young, *The Complaint, or Night Thoughts on Life, Death, and Immortality*, in *Eighteenth Century Poetry and Prose*, 2nd ed., Ronald Press (New York 1956), p. 513.
8. Cited in Lovejoy, *The Great Chain of Being*, p. 190.
9. Lovejoy, p. 190.
10. Will and Ariel Durant, *The Age of Voltaire*, Simon & Schuster (New York 1965), p. 570.
11. Georges Louis Leclerc de Buffon, *Histoire naturelle generale et particuliere*, cited in *The Age of Voltaire*, p. 571.
12. Ibid.
13. Pierre Louis Moreau de Maupertuis, *Systeme de la nature*, cited in *The Age of Voltaire*, p. 578.
14. As paraphrased by Will and Ariel Durant, *The Age of Voltaire*, p. 150.
15. Erasmus Darwin, *Zoonomia or the Laws of Organic Life* (2 vols., London 1794-1796), and Erasmus Darwin, *The Botanic Garden* (London 1791). E. Darwin developed his evolutionary ideas chiefly in volume 1 of *Zoonomia*, which Charles read.
16. *Cassell's Latin Dictionary*, Cassell & Co. (London 1959), p. 393.
17. *Poetry of the Anti-Jacobin*, G. Willis (London 1854) p. 247.
18. See "Darwin, Erasmus," in *The Encyclopedia of Philosophy*, Vols. 1 & 2, Macmillan (New York 1967), p. 295; see also "On Charles Darwin and His

Grandfather Dr. Erasmus Darwin" in Appendix of *The Autobiography of Charles Darwin*, ed. by Nora Barlow, Norton (New York 1958), pp. 150-1.

19. *The Autobiography of Charles Darwin*, p. 94. An interesting sidelight is that Darwin once intended to enter the clergy, but this intention "died a natural death when, on leaving Cambridge, I joined the Beagle as a Naturalist" (*Autobiography*, p. 57). Judging from the *Autobiography*, Darwin's interest in religion was always dilatory.

20. Ibid., p. 93.

21. Cited in Lovejoy, *The Great Chain of Being*, p. 255. Lovejoy cites Maupertuis as saying that nature as we know it is like a once regular edifice after it has been struck by lightning, and he quotes Maupertuis: "it presents to our eyes only ruins in which we can no longer discern the symmetry of the parts nor the design of the architect." Reflecting uniformitarian bias, Lovejoy refers to the Maupertuis idea as a "far-fetched conjecture ...unsatisfactory . . . arbitrary. . ."22. See *Collected Works of Edgar Allan Poe*, Vol. II, Belknap Press (Cambridge 1978), "The Conversation of Eiros and Charmion," pp. 455-61, and pp. 452-3 for reference to Austin's "The Comet."

23. *The Age of Voltaire*, passim, especially Chap. XVI, "The Scientific Advance"; and *The Great Chain of Being*, passim, especially Chapter IX, "The Temporalizing of the Chain of Being."

24. Alfred, Lord Tennyson, *In Memorium*, cited from *Poems of Tennyson*, ed. by Jerome H. Buckley, The Riverside Press (Cambridge 1958), p. 208.

25. *The Great Chain of Being*, p. 236.

26. Carl Bode, "Introduction: Barnum Uncloaked," in P. T. Barnum, *Struggles and Triumphs: or, Forty Years' Recollections of P. T. Barnum Written by Himself*, Viking Penguin (New York 1981, for "Introduction"), p. 12.

27. See *The Autobiography of Charles Darwin*, pp. 121-2 for Darwin's own (accurate) account of the joint publication of his and Wallace's papers.

28. *The Encyclopedia of Philosophy*, Vols. 1 & 2, p. 294.

29. William D. Stansfield, *The Science of Evolution*, Macmillan (New York 1977), p. 9.

30. Immanuel Velikovsky, *Earth in Upheaval*, Doubleday (Garden City 1955), p. 242.

31. Darwin also encountered powerful scientific opposition, e.g., geologist Adam Sedgwick (1785-1873), who wrote that man is a barrier to "any supposition of zoological continuity -- and utterly unaccounted for by what we have any right to call the laws of nature" (see *The Encyclopedia of Philosophy*, Vols. 1 & 2, p. 303). Sedgwick was one of Darwin's teachers at Cambridge.

32. Morris Kline, *Mathematics in Western Culture*, Oxford University Press (London 1953), p. 233.

33. Charles R. Pellegrino and Jesse A. Stoff, *Darwin's Universe: Origins and Crises in the History of Life*, Van Nostrand Reinhold (New York 1983), pp. 151-5.

34. Ibid., p. 155 (emphasis added).

35. Rose, op. cit., p. 39.

36. Charles Darwin, *Origin of Species*, "Recapitulation and Conclusion."

37. Kline, p. 232.

38. In the November 1989 issue of *Natural History*, an article entitled "Birth of the Moon" accuses Velikovsky of "pseudo-science," p. 74. (This article is examined in Chapter IX.)

39. Velikovsky, *Worlds in Collision*, Doubleday (Garden City 1950), passim.

40. Velikovsky, *Stargazers and Gravediggers*, William Morrow (New York 1983), p. 295.

41. This was announced by Velikovsky in a lecture at Princeton University on Oct. 14, 1953, and subsequently communicated to Einstein during the summer of 1954. It was stated in an announcement by Dr. Bernard F. Burke and Dr. Kenneth L. Franklin on April 5, 1955, in a meeting of the American Astronomical Society, that these noises were totally unexpected and unexplained.

42. See Thomas Ferté, "A Record of Success," *Pensée* II-2 (May 1972), pp. 11-15, 23, for a now incomplete enumeration of Velikovsky's verified predictions.

43. Ralph Juergens, "Minds in Chaos," *American Behavioral Scientist* (September 1963), p. 7.

44. Alfred de Grazia, Ralph Juergens, Livio Stecchini, *The Velikovsky Affair*, University Books (New Hyde Park, N.Y. 1966).

45. Kline, pp. 123-4; *Encyclopaedia Britannica*, 1971 edition, Vol. 9, "Galileo Galilei," p. 1,089; *The Encyclopedia of Philosophy*, Vols. 3 & 4, "Galileo Galilei," pp. 262-3.

46. P. Davis and E. Solomon, *The World of Biology*, McGraw-Hill (New York 1974), p. 414.

47. Harold L. Levin, *The Earth Through Time*, Saunders College Publishing (Philadelphia 1983), p. 16.

48. Stephen Talbott, "Impregnable Orthodoxy," *Research Communications Network* Newsletter No. 2 (February 10, 1977), p. 1.

49. Ibid., pp. 1-2.

50. Ibid.

51. Kline, p. 231.

52. Ibid., p. 233.

53. Alexander Rosenberg, *The Structure of Biological Science*, Cambridge University Press (Cambridge 1985), pp. 236-7 (emphasis added).

54. William D. Thornbury, *Principles of Geomorphology*, Wiley (New York 1969), p. 17.

55. *Encyclopaedia Britannica*, 15th ed., "Stratigraphic Boundaries," Vol. 17, p. 17.

56. Windsor Chorlton, *Ice Ages*, Time-Life (New York 1983), p. 85.

57. Diane P. Gifford, "Taphonomy and Paleoecology: A Critical Review of Archaeology's Sister Disciplines," *Advances in Archaeological Method and Theory*, ed. by M. B. Schiffer, p. 389 (emphasis added).

58. Thornbury, op. cit., p. 391 (emphasis added).

59. Dwardu Cardona, "Let There Be Light," *KRONOS* III-3 (Spring 1978), p. 48.

60. Bennison Gray, "The Science of Evolution," *KRONOS* VIII-3 (Spring 1983), pp. 40-1.

CHAPTER II

The Creationists: Rebels Without a Cause?

Darwinism has been subjected to much critical scrutiny, no small part of this by creationists. Certainly a large part of the published criticism has appeared in creationist literature. Typically, if this material is addressed at all by evolutionists, the challenge to Darwinism is evaded by recourse to ridicule of Creationism. Creationist proposals are viewed as sublime nonsense -- but mostly for subjective reasons.

Amid the polemics, there has appeared no real analysis of the creationist position relative to its foundations. If examination could show that disqualifying contradictions and other faults cling to Creationism, and if the Darwinists would recognize the sterility of their own dogma, perhaps the ridicule would cease and all interested parties might focus their energies on a redirected search for truth.

I have undertaken the study presented in this chapter as an analyst constrained to the English language. Consequently, the entire chapter depends on the scholarship of others in its evaluation of the Old Testament record, originally transcribed in Hebrew. Specifically, it rests on the most widely accepted English translation of the Hebrew record even though scholarship has shown that the Genesis account of beginnings is radically altered when correct translations of basic terms are used.[1] The text read by millions of Christians is the right one for this chapter because its errors are Creationism's errors.

Creationist viewpoints lean, for the most part, on the Genesis text most widely current in English (although fine points of the Hebrew text occasionally are debated). If creationist interpretations of the King James rendering of Genesis are negated by logical analysis, those interpretations can be safely rejected without recourse to the true, complete meanings of the Hebrew texts. It is inconceivable that a revised interpretation of the Hebrew sources would be able to restore the creationist framework; its rejection would stand. In short, this chapter challenges the creationist position on its own terms, from its own perspectives.

The Biblical Record

The biblical creation account appears in the first two chapters of the Book of Genesis. One version is found in Genesis 1:1-2:3, another in Genesis 2:4-23. Some commentators maintain that the two chapters are independent (written by different authors) and, in fact, somewhat contradictory; others contend that they actually are complementary. Chapter 1 provides most of the details that we wish to examine, describing a six-day creation process. During this time frame, God (Elohim, plural) calls into being basic features of the earth and all of its living creatures.

In the "beginning," God created the heavens and the earth, and on the first day He made light, night, and day. Creationists interpret "light" to mean all wavelengths of the electromagnetic spectrum.

On the second day, the earth's waters were divided concentrically -- above and below the "firmament." The significance and ramifications of such a configuration are addressed below.

The third day witnessed the appearance of dry land and plants, and the fourth brought forth the Greater and Lesser Lights and stars. Creationists circumvent the difficulty of light preceding the sun by arguing that the sun was obscured by atmospheric vapors until the fourth day, when the vapors cleared and the sun made its first appearance. But such reasoning would require a terrestrial observer -- which, in the creationist scheme, did not yet exist.

On the fifth day came water dwellers and birds, and on the sixth, land dwellers, including man.

The acts of creation are viewed as the winding-up of the universe, as of a watch; natural law did not apply. There was necessarily an appearance of age: soil capable of supporting plant life; animals capable of reproduction. Moreover, "radioactive clocks" might have been "set" at any stage of decay. In such a context, no question on origins is unanswerable; the imagination can range at will, making and unmaking rules as may seem needful. The creationist is as fruitful of explanations as is the uniformitarian.

Adam and Eve are introduced in the second account of beginnings (Genesis 2:4-23). The first account is less specific than the second about man's new presence; Genesis 2 specifies that Adam was created from the dust of the earth.

Clearly, many of the Creation elements will always be challenged by skeptics as incredible. Except to consider one other question, however, this chapter will address only the apparent fallacies in creationist interpretation and logic in relation to Creationism's accepted English text of Genesis.

The one other question: Is "Scientific Creationism" really scientific?

The philosopher of science has to respond with a resounding "No." In order for a proposition to qualify as scientific, it must be testable in principle. That is, there must be the possibility of proposing a test that could either confirm or disprove the proposition (actually conducting the test is not necessary).[2] Obviously, no such test could be devised to determine the truth or falsity of Creationism. It follows that Scientific Creationism is not scientific.

But that might just be the least of its problems.

The Genesis "Days"

A central problem in the interpretation of the first chapter is the meaning of the Hebrew *yom*, usually translated day. The problem has arisen as a result of recent teachings from historical geology that the earth is some 4.5 billion years old. These teachings have led to a variety of opinions and interpretations among biblical scholars.

The most popular interpretation is that the creation days were ordinary solar days, maybe or maybe not twenty-four hours in length.

Then there is the "day-age" theory, suggesting that the days were possibly great eons of time. This theory is an obvious attempt to preserve Scriptural integrity in an evolutionary-minded world.

Another variation makes each of the days 7,000 years long. This interpretation is based on a number of passages from other parts of the Bible having no direct relation to the creation issue.

Probably one of the least popular interpretations is that the days are merely indicative of days on which Moses (the presumed author of Genesis) received visions of the creation activities.

Some creationists believe that between the first two verses of Genesis there could have been a period of time of any length, from thousands to millions of years, and that this period could include what has come to be known as "the geological ages." After this interval of undetermined length, Elohim re-created the earth, making it anew in a week of solar days.

The probable reason for the popularity of this belief is the exposure provided for it in the Scofield Reference Bible, widely used by conservative Christians and Christian institutions. R. F. Surburg correctly observes that "this theory cannot be substantiated from the Bible. It attempts to give some explanation for the different strata which, geologists say, make up the surface of the earth. But the gap theory gives no explanation for the fossils in the rocks unless, as Berkhof says, 'It is assumed that there were also successive creations of animals, followed by mass destruction.'"[3]

Surburg goes on to enumerate seven reasons why "[d]espite [the] many variant ideas concerning the meaning of 'yom,' the 'days' referred to in connection with the creative activity of God were not long periods of time but solar days of approximately 24 hours."[4]

It is the literalists (cf. 24-hour day) who are at the forefront of the creation-evolution controversy. To evaluate basic literalist concepts, one must examine their supposed biblical underpinnings.

The idea of a literal day is clear enough. It comes from a straightforward reading of the English text. Each day is said to have an evening and a morning, darkness and daylight.

A Young Earth?

The accepted literalist time of the Creation, some 6,000 years ago, may be less obvious. This figure is based upon genealogies of the Hebrew patriarchs recorded in Chapters 5, 10 and 11 of Genesis. Lifespans are provided, with the number of years of father-son overlap. Descent is traced from the first man, Adam (not identified in the first version of the Creation), to Abraham. Abraham can be roughly dated, and counting backwards from him puts Adam at around 4004 B.C. This date apparently was first derived by Bishop Ussher, a respected scholar in the seventeenth century. Many Bibles continue today to perpetuate this date, which they carry in their upper page margins. Figures are different, however, in the Septuagint, an ancient Greek version of the Old Testament. Moreover, the Septuagint includes an additional name, which also is found in a recapitulation of genealogies in the Book of Luke (3:36; Cainan). Then, too, there is some debate among creationists as to the completeness (continuity) of the genealogies.[5]

The Genesis "Kinds"

Explicit in the creationist concept, and in opposition to evolution, is the fixity of species. This concept does not necessarily refer to existing scientific classifications of species; rather, it embraces *natural* species. A natural species is usually considered to be any grouping of organisms that is interfertile, i.e., capable of interbreeding. In this sense, dogs and wolves are part of the same species. This conception of species might be extended to include hybrids, such as mules.

A natural species is commonly alleged to be equivalent to the biblical "kind" (Hebrew *min*) divulged in Genesis. The central pillar of creationist thought is that all plants and animals reproduce "after their kind"; one kind cannot produce another kind. This fundamental concept appears with varying emphasis throughout creationist literature.

Writes Surburg: "An important matter brought out clearly by the Genesis account is that the author places a limit on change in living things. In vv. 21, 24, and 25 the Mosaic account of Chapter 1 states that plants and animals reproduce after their kind. . . . The expression 'after its kind' would then mean that Jehovah made plants and animals according to their various divisions. It means that these are definite limits beyond which plants and animals may not vary."[6]

Byron Nelson elevated this to a principle: "After its kind is the statement of a biological principle that no human observation has ever known to fail."[7]

F. L. Marsh hoisted the idea up another notch, making it a law: "[T]he assertion of Genesis is that each [organism] continues to bring forth after its kind . . . [and] the only type of descent which Genesis describes was still the rule of reproduction in Noah's day -- each basic type of plant continued through the generations true to the law of its creation and brought forth after its kind."[8]

In addition to kinds, the Book of Genesis speaks of created groups, such as cattle, fowl, etc. If fertility were not the criterion for species definition, one might ask whether kinds exist within the fowl group or whether, perhaps, multiple fowl groups exist within a kind. All such questions turn out to be academic, however, since a close reading of the biblical text reveals that the accepted meaning of "kind" is unsubstantiated and that the true meaning is far from being clear.

Consider Genesis 1:25: "And God made the beast of the earth after his kind, and cattle after their kind, and every thing that creepeth upon the earth after his kind. . . ." The creatures were *made* after their kinds. Nothing is said about subsequently *propagating* after their kinds.

The statement about plant life in Genesis 1:11 seems at first to uphold the creationist contention about "propagation after its kind": "And God said, Let the earth bring forth grass, the herb yielding seed, and the fruit tree yielding fruit after his kind, whose seed is in itself. ." Unlike the sense adduced for the previously cited verse, the meaning perceived by creationists in this one is strongly influenced by the sentence structure and punctuation of the King James translation. For compatibility with the Hebrew texts, it must be read as follows: "Let the earth bring forth grass, the herb (the seed yielder), and the fruit tree (the fruit yielder), after his kind, whose seed is in itself."[9]

The earth brought them forth after their kind: This, again, is a creative act and does not imply reproduction from a parent organism. This reading finds further emphasis through the absence of the word *min* from Genesis 1:27: "So God created man in his own image, in the image of God created he him; male and female created he them." Like every other "creature" (Hebrew *nephesh*, usually translated "soul," not "creature" as in King James/Genesis 1:24), mankind does procreate, but the emphasis of the statement -- just as in the other statements, if they are read aright -- is on the pattern after which God created the new life.

"Kind" is therefore not a biological entity; rather, it appears to represent a concept, a plan, apparently envisaged as being in the mind of God, after which plants and animals were made. No restrictions are imposed on biological variations or development. The writer was interpreting the order of things, not describing a law of nature.

(It is interesting that the idea apparently intended by the Genesis author seems to be exactly equivalent to Plato's later concept of Forms -- the Form being the ideal pre-existent model reflected in all individuals patterned upon it. The ultimate form was God himself, in whose image man was made.)[10]

Evolution cannot be challenged on the basis of immutable Genesis "kinds."

The Vapor Canopy

Another integral part of creationist belief is the concept of a primordial vapor canopy, probably first proposed by Isaac Vail over a century ago. It is alleged that in the ancient past, a shell of water vapor surrounded the earth at some unknown distance from the surface. This idea is so popular and so ingrained that one can hardly find an issue of the *Creation Research Society Quarterly*, the leading creationist journal, that does not contain a reference to it.

The basis for the canopy is found in Genesis 1:7: "And God made the firmament, and divided the waters which were under the firmament from the waters which were above the firmament. . . . "

Henry Morris has promoted the idea of the primordial vapor canopy in many of his publications. In his words:

> The waters "above the firmament" seem to imply more than our present clouds and atmospheric water vapor, especially since Genesis 2:5 implies that during this time rainfall was not experienced on the earth.
>
> Although we can as yet point to no definite scientific verification of this pristine vapor protective envelope around the earth, neither does there appear to be any inherent physical difficulty in the hypothesis of its existence, and it does suffice to explain a broad spectrum of phenomena both geological and Scriptural.[11]

Morris also extols the effects of the canopy on meteorological and hydrological activities and cycles, as well as on terrestrial life, painting an inviting picture of the ancient world: "[The] canopy of water vapor . . . provided a warm, pleasant, presumably healthful environment throughout the world. Perhaps the most important effect of the canopy was the shielding action provided against the intense radiations impinging upon the earth from space."[12]

No such canopy exists now, of course, and its presumed catastrophic collapse provides another explanation readily accepted by creationists. Morris theorizes that "if we accept the Biblical testimony concerning an antediluvian canopy of waters . . . we have an adequate source for the waters of a universal Flood."[13]

The Flood will be addressed more fully in later chapters, but for now, a glance beyond the single verse on which the canopy idea is based exposes a fallacy. The definition of firmament is implied in Genesis 1:20: ". . . fowl that may fly above the earth in the open firmament of heaven." The firmament thus includes the lower atmosphere. It is referenced again in verse 17, where it is stated that the Greater and Lesser lights were placed in it.

Morris and J. C. Whitcomb, Jr. describe the firmament as "an expanse of atmosphere in which birds were to fly (Gen. 1:20) and in which light from the sun, moon* and stars was to be refracted and diffused to give light on the earth." (Gen. 1:17)[14] But they fail to take into account Genesis 1:14: "Let there be lights in the firmament of the heaven to divide the day from the night; and let them be for signs . . ." The designation as signs makes clear the fact that light sources (in the firmament) are meant, as opposed to refracted and diffused light coming from beyond, or outside, the firmament, from whatever sources. Light that was "refracted and diffused" through a "canopy of waters" would not permit the discernment of individual light sources, and thus the sources ("lights in the firmament") could not be used "for signs."

The Greater Light and Lesser Light were placed in the firmament, but the waters were above it. If the waters above the firmament constituted a shell of water vapor around the earth, the two lights would have been enclosed as well -- an impossible configuration.

Thus, the firmament was not synonymous with the earth's lower atmosphere, and the waters above it are not properly to be described as a stratospheric vapor canopy.

When I first published my observations on this matter, they were greeted rather defensively in a creationist newsletter. According to the newsletter editor, "It is possible for the word 'firmament' to have two meanings. . . . In one instance firmament may refer to the lower atmosphere in which the fowl fly [Genesis] 1:20, while in [Genesis] 1:17 the reference may be to space where God placed the sun, moon, and stars."[15]

What can be said to such fast-and-loose parrying with the word "may"? Any condition "may" have existed if one is content to abandon

---

* As to the moon, however, see Chapter IX.

consistency and logic. Whatever the true meaning (or meanings) of the word "firmament," there is no biblical accommodation for any kind of canopy. An objective basis for a definition does not exist, so a single verse of Scripture is tortured to support a belief that not only has no foundation but also is contradicted by other Scripture.

Conclusions of Faith

So we see that in addition to not being scientific, "scientific/biblical" Creationism is not entirely biblical, either. Major points in the creationist position, supposedly based on biblical pronouncements, actually have no biblical basis. Moreover, since the biblical creation account itself purportedly covers the time before man's presence, its acceptance as history by the creationists is strictly a matter of faith. Despite the supposed scientific basis for Creationism, the anchorage in faith often is acknowledged. Writes John W. Klotz:

> I believe that we must recognize that our point of view is based ultimately on faith. It is a mistake to base creationism on empirical observations. The very nature of the problem does not lend itself to this, since the experimental method, the genius of modern science, has very little use in the study of evolution. Our opposition to evolution is based on our acceptance of the Bible as God's Word. We believe that the Genesis account is a factual account.[16]

And, indeed, much of it is. But an interpretation forced to conform to a particular system of belief makes it of little value to our understanding and enlightenment, and there are other weaknesses in the creationist position to be analyzed hereafter.

The word "scientific" as applied to Creationism seems to be limited to the methodology of creationist attacks (outstandingly devastating) on Darwinism. But even if Darwinism were refuted to the total satisfaction of all concerned, nothing would be contributed thereby to substantiating the Book of Genesis or, most especially, the creationists' faulty interpretation of it.

On the other side of this coin, even if the Genesis record of beginnings could be shown to be a complete fabrication, conscious and directed creation would not be excluded from the realm of possibilities.

The creationists' argument for a young Earth tends to cloud the issue. Despite its being grounded in alleged historical data, this argument must nevertheless assume (i.e., accept on faith) an "end-of-creation" starting point. Proving the earth to be only a few thousand years old would sound the death knell for Darwinism, but it would establish nothing about the factuality of Divine Creation or the Genesis account of it. Creation of the world and the concept of a young Earth are completely independent of each other.

In the mazes of Creationism, objectivity is a will-o'-the-wisp. Then again, perhaps no one really expects biblical literalists to be other than subjective. Baffled expectations are more often the result of running headlong into the mist of question-begging subjectivity that is the preserve of Neo-Darwinism.

REFERENCES

1. Dwardu Cardona to J. E. Strickling, personal correspondence (January 1985).
2. Carl G. Hempel, *Philosophy of Natural Science*, Prentice-Hall (Englewood Cliffs 1966), p. 30.
3. R. F. Surburg, "In the Beginning God Created," from *Darwinism, Evolution, and Creation*, ed. by P. A. Zimmerman, Concordia (St. Louis 1959), p. 53.
4. Ibid., p. 59.
5. See discussion by H. M. Morris and J. C. Whitcomb, Jr., *The Genesis Flood*, Presbyterian and Reformed Publishing Co. (Philadelphia 1961), pp. 481-3.
6. Surburg, op. cit., pp. 64, 67.
7. Byron Nelson, *After Its Kind*, rev. ed., Bethany (Minneapolis 1967), p. 3.
8. F. L. Marsh, *Life, Man, and Time*, Outdoor Pictures (Escondido 1967), pp. 144, 135.
9. My paraphrase.
10. W. K. C. Guthrie, *The Greek Philosophers*, Harper and Row (New York 1950), p. 88.
11. Morris and Whitcomb, op. cit., pp. 215, 241.
12. Ibid., p. 399.
13. Ibid., p. 77.
14. Ibid., p. 215.
15. *The Bible-Science Newsletter* XIV-8 (June 1978), p. 8.
16. John W. Klotz, "Creationist Viewpoints," from *A Symposium on Creation*, Baker (Ann Arbor 1968), pp. 48-9.

CHAPTER III

Darwinism: a Barrier to Evolution

What is the meaning of *evolution*? And what is the substance of the evolutionist position?

The accepted fundamental of evolution is more or less recognized by the general public. To wit, all living organisms have developed over long periods of time, entirely by chance, from other organisms that were of a "lower order." This is the view of most (but not all) evolutionists today.

The orthodox conception (current scientific view) of the process of evolution -- mutation and natural selection -- is variously known as Neo-Darwinism, the Modern Synthesis, or the Synthetic Theory. It represents the integration of genetic theory into evolutionary biology.

Ongoing reinforcement of evolutionist concepts is assured through our educational institutions. If evolution is taught at all in high schools and colleges, it typically is based on textbooks that portray Neo-Darwinism as a demonstrable scientific fact. Meanwhile, Creationism usually is attacked as nonsense unworthy of scientific scrutiny.

Creationists, as noted in the previous chapter, will admit their reliance on faith for the positions they espouse. But a Diogenes wandering the earth with his lantern would be hard put to find an evolutionist who would acknowledge the vast amounts of speculation underlying evolutionary theory.

To the contrary, we are informed by Richard Dawkins, author of *The Selfish Gene*, that "Darwin's theory is now supported by all the available relevant evidence, and its truth is not doubted by any serious biologist. . . . I suggest that it may be possible to show that, regardless of evidence, Darwinian natural selection is the only force we know that could, in principle, do the job of explaining the existence of organized and adaptive complexity."[1]

And from *Encyclopaedia Britannica*: "It is [natural] selection . . . that controls the direction, rate, and intensity of evolution. This conclusion, based on experiments in genetics, is *confirmed* from work in paleontology."[2]

Behind closed doors, however, we perhaps find more marketing than objective analysis of facts. In a review of *The Evolutionary Synthesis: Perspectives on the Unification of Biology*, Niles Eldredge reveals that a few prominent and persuasive leaders are behind the hegemony of the Synthesis (Neo-Darwinism). The vast majority of biologists have accepted the Synthesis, Eldredge writes, because of "the persuasiveness of a few highly talented biologists, promulgating a single, simple, and rationally very appealing set of ideas."[3] But widespread agreement is not proof -- it is, in fact, no more than consensus. And consensus is hardly synonymous with the confirmation proclaimed in *Encyclopaedia Britannica*.

The confusion of thought goes deeper than many scientists are willing to recognize. In Synthesis literature there is no apparent distinction between Darwinism and evolution. According to scholar Norman MacBeth: "The evolutionists themselves forget that there is a difference between evolution and Darwinism and ascribe everything to natural selection just as though it were equivalent to change. This has caused untold difficulty. Darwin himself, of course, was continually confusing them [i.e., confounding evolution and natural selection]. . . . In the common mind of the Western World they are indistinguishable."[4]

The air needs to be cleared. To prove evolution is *not* to prove Darwinism (although proving Darwinism *as generally perceived* would make a case for evolution.).

Several questions can be asked of Darwinism. What is the true nature of "natural selection"? Is it only a hypothesis? Or if evolution is an actual process of nature, is it truly to be explained by natural selection? If evolution is not "selection," what is it?

For that matter, what precisely is a biological *species*, the presumed focal point of evolutionary process? Any inquiry into the validity of Darwinism must begin with an overview of the conceptual problems surrounding both species and the nature of speciation.

Background[5,6,7,8]

Man's preoccupation with recognizing biological species is rooted in survival, from antiquity to the present day, because it is important to know which species are useful to man, and for what purposes, and which ones are harmful, and in what ways.

An informal basis for species recognition is sometimes sufficient for small regions or small groups of people, but as scientists began attempting to name and classify all living and fossil species, a more universal system of nomenclature became essential for scientific communication, as did a system based on relationships among species.

John Ray, a seventeenth century English naturalist and one of the first men to suggest a formal arrangement of species, proposed that species be recognized on the basis of structural differences inherited by one generation from another. The system in use today is a modification of one originally developed by Swedish naturalist Carolus Linnaeus. It features a generic and a specific name for each known organism in nature, both plant and animal.

Linnaeus believed that each species was the result of a separate act of creation and never changed. It followed that a simple individual organism could characterize an entire species. Through the eighteenth century, the term "related species" generally meant to biologists that species simply shared certain characteristics. Biologists usually looked for a single principle as a basis on which any two groups could be defined as a single species or as two separate species. For some, this principle was reproductive fertility (ability to interbreed); for others, it was morphology (physical form); and for still others, various adaptations or ecological roles.

With the speculations of such eighteenth century scientists and philosophers as Maupertuis, Buffon, Lord Monboddo, Hutton, Diderot and others, the term "related" began to acquire a new meaning, one that had penetrated popular thought sufficiently by 1842 for P. T. Barnum to make a lot of money by displaying fake "missing links" between man and other species. The new meaning of "related" was "evolutionary descent from a common ancestor." With the publication of Charles Darwin's *On the Origin of Species* (1859), the new concept was given an overpowering impetus among the scientific community, although simultaneously popular opinion was mobilized against it (evidently because Darwin raised the concept of evolution from a curiosity to a serious threat to conventional beliefs).

Upon the rediscovery in 1900 of Gregor Mendel's earlier work on heredity (related below), species differences came to be understood as characteristics inherited by means of the now-familiar genetic mechanism. A species is recognized today as a population of organisms that share a number of traits (based on genes) which appear in the offspring resulting from matings of individuals of that population. A more precise definition thus far has proved to be problematical. Concepts vary. In fact, there is at present no general agreement in biology on the explicit definition of the term "species." And most of the current definitions, together with the theories of speciation in which they figure, are incompatible with each other.

Some biologists contend that a species names a group of organisms capable of interbreeding and separated from other species by a lack of this ability. Others see it as a class constrained by evolutionary or ecological forces. Physical boundaries to the association of populations (gene flow) also are involved in definitions. One extreme rejects the very notion that species exist as such, insisting instead that a particular species is no more than a collection of particular biological individuals scattered through space and time. Such divergence of opinion has enormous ramifications for evolutionary theory.

Darwin did not solve the problem of the origin and diversity of species; he was not even sure of the meaning of "species." Over a hundred years later, there now is less agreement on an explicit definition than there was in Darwin's day. Neither is there a consensus on what counts as a sufficient condition for species membership, or for what counts as a mark of that membership. There is no universal specification or description of properties or kinds of properties for determining whether an organism belongs to a given species. And without a solution to this definitional problem, there is no distinctive Darwinian theory of natural selection.

Today, biology is dominated by three prominent perspectives ("definitions") of species and associated accounts of speciation (only one of which is addressed at length here): biological; evolutionary; and an ecological treatment of species. All three reject the idea that a particular species can be given an explicit (or nearly so) definition. This posture is in opposition to the notion of *biological essentialism*, which holds that for each species there is a set of characteristics belonging to its individual organisms that sets them apart from other species, thereby resulting in exclusive membership for their own species. Essentialism has been held to be the chief intellectual obstacle to the idea of speciation by natural selection.

From before Plato through most of the nineteenth century, the conception of species was essentialist. Species were viewed as elements were; the members of each species were seen as having properties in common. In the absence of evolutionary theory, these shared properties explained a group's place and distinctiveness in nature. However, the entrenchment of this essentialist conviction came to be seen as an impediment to evolutionary thinking. It made discussion of the evolution of species difficult. The essentialist approach thus is rejected in the framework of natural selection; it typically is viewed as a conceptual mistake to be avoided at all costs.

What, then, is a species? Can it be said to have a real existence? Benjamin Burma (*Evolution*, 3:369) muses that a species would seem to be all of the generations through time of a line of interbreeding creatures. But on the evolutionary assumption that there is a biological continuity from all life back to the "original primordial cell," all life then would belong to one single species. On the other hand, if the species is defined as all of that same group at a single point in time, the group as constituted is only one in an infinite series of species, since the time through which the group's ancestors lived is infinitely divisible. No wonder, then, that some scientists see species as having an existence only in our minds.

But others disagree.

Most biologists think of species in terms of a biological, or *natural*, species. This is a concept allowing that a single species may be spread over a wide area and that in different areas it may be very dissimilar, either in appearance or in behavior. The most widely adopted version of this concept of species comes from Ernst Mayr. As originally postulated by Mayr, a species was defined as a population of organisms capable of interbreeding, whether or not given the opportunity, and reproductively isolated from other organisms. But from a theoretical point of view, the notion of reproductive isolation is actually indeterminate, i.e., one could not predict in advance whether two groups could or could not interbreed. Moreover, some populations identified as belonging to different species can hybridize. Whenever this happens, Mayr's concept of biological species breaks down.

A major theoretical problem is the reconciliation of the concept of species as "reproductively isolated interbreeding populations" with the existence of asexual species. Asexual species reproduce themselves without interbreeding, without exchanging genes. Their members are as completely isolated from one another as they are from members of any other species. This problem is readily recognized but typically is played down as a minor annoyance.

Mayr concluded that *sexual* species may be defined in terms of interbreeding and *asexual* ones in terms of morphology (physical structure). Indeed, these are useful indicators of species differences. Even so, the first of these will not identify different sexual species if subspecies or separate groups of a given species do not have the opportunity to interbreed. For asexual species, morphological differences would merely be marks of membership in a particular group, a group that could not satisfy the interbreeding-based definition of biological species.

But both interbreeding and morphological differences could be seen as joint effects of some other single determinant, variations of which could be cited in a definition of species applicable to both sexual and asexual taxonomy. Mayr justified his appeal to morphology on the ground that genetic difference is responsible for both it and reproductive isolation. Reproductive isolation then would not be a cause of diversity; rather, it and morphological differences would be mere *signs* of diversity. The cause would be genetic variance, which would supplant interbreeding or morphology as a basis for definition.

Further difficulties for the notion of biological species arise from the fossil record. One means of speciation is presumed to be *anagenesis*, that is, the evolution through one line of descent of a single species into another, separate species (Gr. *ana*, "again," and Gr. *genesis*, "birth"; thus, "birth again"). Each organism in the line is reproductively connected to its immediate ancestors, but the organisms at the extremes of this evolutionary line (the beginning and the end) supposedly are of different species. *Cladogenesis*, the separation of a genealogical line into two separate species, faces many of the same conceptual problems as its sister concept. (Cladogenesis is from Greek *klados*, "a shoot, branch," plus *genesis*; thus, it is a "branched birth.")

Gradual anagenesis is obviously incompatible with speciation and current notions of biological species; succeeding generations can always interbreed, so a species cannot be delimited within "a slice of time." This means that populations at the end of a given lineage are not isolated from those at its beginning. We must deny the occurrence of anagenesis or else deny that it is ever gradual. (We shall see that the so-called "punctuated equilibria" approach does not reject gradualism as its adherents maintain; it merely speeds it up. Thus it offers no solution to the conceptual problem.) We are therefore forced to choose whether there is anagenesis and whether evolution is gradual. Acceptance of the reality of biological species and anagenesis denies gradualism; arrival of a new species would have to be sudden.

In the case of asexual species, what actually counts as speciation in their evolution is completely indeterminate. As noted above, the notion of reproductive isolation as the cause of speciation cannot be reconciled with the existence of asexual species. All asexual genealogical lines are going their separate ways regardless of proximity to or remoteness from one another. To appeal to morphological differences, however, is to change the standard of species definition. To establish morphological variance as a determinant of species differences is to reject a biological definition of species.

Until the two definitions (interbreeding and morphology) can be explained by a theory that supports both, it must be concluded that the definitions do not have the generality required to explain species diversity and its causes. So far, Mayr's notions have done Neo-Darwinism no good; Neo-Darwinism has not been able to develop a theory that supports *either* of Mayr's definitions.

A Conceptual Failure

So after more than a century of research, science has no clear understanding of the nature of species. Popular accounts are inadequate to explain what a species is. If species do not constitute natural kinds, the Neo-Darwinian is hard put to prove that there is any single type of event called speciation or that there exist single types of organisms called species. Yet orthodox science remains convinced that a species is "something" produced by natural selection -- "something," but not a *natural kind*, which would invalidate Neo-Darwinism. And natural selection has proven to be just as great a failure in explaining "the origin of species" as have definitions of species.

Anagenesis and cladogenesis prove to be on a collision course over the issue of natural kinds. If the notion of a species as a natural kind is retained, what place is there for gradualistic anagenesis (natural selection)? Where does one natural kind leave off and another begin in such a progression?

On the other hand, if the notion of species as a natural kind is discarded, what place is there for the concept of cladogenesis? For one species to split from another, there must be some kind of division or difference to bring it about. Division or difference from what? From a mere collection of individuals at a point in time? How is it to be determined that the "branched birth" of a "new" population has occurred? The interbreeding test? Inability to interbreed with the "old" population would annul the possibility of gradualism. The morphological test? If the "new" were able to interbreed with the "old," morphology would not be conclusive in itself. A combination of tests? At this point, cladogenesis faces all of the objections just adduced against gradual anagenesis.

We are told by Mayr that two populations must become reproductively isolated from each other before speciation can begin. If that is so, exactly when is cladogenesis supposed to occur? Mayr cannot have it both ways, commingling and isolated. To "branch off" by means of a genetic change, does the "new" not have to issue from a commingling of the "old"? If by "branching off" or "branched birth" is meant simply the development of separate species in reproductive isolation, what is the difference between the processes of cladogenesis and anagenesis? Geographical separation would serve only to put the groups on separate paths of evolutionary descent (i.e., would merely set anagenesis in process). (From a strictly orthodox-definitional point of view, cladogenesis merely implies the continued identifiability of the "parent species" as such. This identity would be lost in the process of anagenesis.)

And what of asexual species, which are reproductively isolated even from their own kind? What has natural selection to do with species such as this? The interfertility basis of species is likewise thwarted.

Admittedly, this discussion of grounds of dissent from Neo-Darwinism has been heavily philosophical. However, if anything is clear from such discussion, it is that current theories are muddled.

Yet one does not have to rely on the rarified air of theory to dash the Neo-Darwinian helium balloon to pieces. Hard evidence from modern science will do the job quite well.

Heredity

The modern science of genetics finds its origins in the work of Gregor Mendel (1822-1884), a monk who spent much of his life studying the common pea plant.

For many years Mendel experimented with several contrasting pairs of clearly defined characteristics of this plant, searching for an understanding of heredity. In so doing, he sought traits that would be either clearly present or completely absent, realizing that if heredity were to be understood, a trait would have to be easily recognizable. His choices of trait were color (red or white), stature (tall or dwarf), and seed shapes (smooth or wrinkled).

When fertilization occurred, Mendel found, three different color combinations were possible, one in two different ways: red-red; red-white or white-red; and white-white. Red is dominant; therefore, three of four plants (on average) will bear red flowers, while one of four will bear white ones.

Thus, when the sex cells of an organism are formed, hereditary factors separate. The sex cells contain either one factor or the other, such as the red or white in Mendel's hybrids, but not both.

Mendel went on to determine that the patterns of inheritance of different traits are independent of each other. Among Mendel's plants, for example, there was found to be no relationship between plant height and flower color.

Mendel published his study in 1866, but it was ignored until 1900, sixteen years after his death. At that time Hugo De Vries, Karl Correns and Erich Tschermak independently rediscovered the same principles. Within the next year, it was demonstrated that they also apply to human heredity. The basic rules were seen to be universal. But the inheritance of a great number of traits does not follow the basic patterns. This divergence occurs because the actual means of inheritance are generally more complex than was implied in Mendel's simple model experiment.

Many genes might come to bear on the same characteristic. One such characteristic in humans is stature. It results from the interaction of some unknown number of genes. Moreover, a single gene might exist in multiple forms (alleles). Instead of the simple red-white encountered by Mendel, there could be red, yellow, white or blue, for example. There also are genes that modify other genes and genes that initiate or block the activity of other genes. And a single allele might affect an entire series of traits. In other words, heredity is a complex, multifaceted process.

Around 1900 scientists began to probe the cell, the basic building block of all life, for the physical and chemical bases of heredity.

When a cell begins to divide, its chromosomes become visible as long rope-like structures within the nucleus. For each species in nature, there is usually a characteristic chromosome number. Humans have one pair of sex chromosomes and twenty-two pairs of non-sex chromosomes, called autosomes. The chromosomes carry the genetic material -- heredity information for perpetuating the organism. The information is copied during cell division.

The possible number of combinations of genes is increased during cell division. This phenomenon involves the exchange of genetic material inherited from the father and the mother. Single chromosomes then carry genetic material from both parents.

The offspring does not necessarily receive equal numbers of genes from all grandparents; of those received from one parent, more might have come from the grandmother than from the grandfather. The number of possible combinations becomes astronomical. Some traits, however, are inherited on the sex chromosome; their patterns of transmission are different from those of autosomes.

Mutations: Chromosomal Abnormalities

Hugo De Vries (1848-1935) succeeded in resolving certain ambiguous concepts concerning the nature of the variation of species. These ambiguities had precluded universal acceptance and active investigation of Darwin's system of organic evolution. De Vries envisaged evolution as a series of abrupt changes radical enough to bring new species into existence in a single leap, a phenomenon for which he coined the term *mutation*. It was during experiments with plant breeding that he drew up the same laws of heredity that Mendel had drawn up. While surveying the literature on the subject of heredity, he discovered Mendel's paper of more than thirty years earlier.

Orthodox science today ascribes much less creative ability to mutations than did De Vries. A mutation is now defined as any alteration of the genetic material. There are two types on the chromosomal level, called chromosomal aberrations. These are abnormal chromosome number and abnormal chromosome structure.

A common error that occurs during division of the sex cells leads to abnormal numbers of chromosomes in the gamete (sperm or ovum). In particular, one extra human chromosome can lead to the condition known as Down's syndrome.

More common are extra or missing sex chromosomes, producing various physical abnormalities and sometimes mental retardation. Moreover, there are several types of chromosome structural abnormalities that can occur.

Genetic material can be lost when a chromosome breaks and a segment fails to be included in the second-generation cell. On the other hand, there can be repetition of a section of a chromosome.

Sometimes parts are reunited in reverse order or else unite with other chromosomes after breaking: Genetic material is retained, but the information it carries is altered.

Abnormal chromosome numbers and structural aberrations are responsible for significant numbers of defects in animal embryos, frequently resulting in spontaneous abortions.

Point Mutations

Genes are segments of chromosomes programmed to perform very specific functions in the reproduction and development processes, as well as other functions throughout the life of the organism. Mutations arise when random changes are made in their code, at a specific point on the DNA molecule. A small error can have extreme consequences.

One such error leads to Phenylketonuria (PKU) in humans, which can cause mental retardation. Other point mutations can produce dwarfism and albinism (lack of pigmentation in the skin and eyes).

Assessment of the Nature of Mutations

So far the roll call of results of genetic alterations has included a number of disorders. What about the other side of the coin? Is one not to expect beneficial mutations as well? Anthropologists P. L. Stein and B. M. Rowe provide an assessment of this issue that displays the convoluted thought underlying Neo-Darwinism:

> A mutation is a mistake . . . whether it is a change at a particular point on the DNA molecule . . . or an aberration of the chromosomes.

> What is the probability of such a chance alteration of the genetic code being advantageous to the organism? Imagine a Shakespearean sonnet being transcribed into Morse code. Suppose that in transcribing the sonnet, a dot was chosen at random and replaced with a dash. What is the probability that this change would improve the poem? Probably it would result in a misspelled word; it might even change the word and, hence, its meaning.
>
> In a like manner, the probability that a chance alteration in the genetic code would be an improvement is very low. In fact, mutations are almost always deleterious to individual organisms. Most result in the death of the zygote or embryo. In addition, since a population is probably adapted to its environment, an individual who has a new mutation will in many cases reproduce fewer offspring in that environment than the other members of the population.[9]

Many biological characteristics cannot be attributed to a single gene, depending instead upon several. Need it be asked what happens to the sonnet when multiple letters are randomly changed at one time?

Sources of Variation

Stein and Rowe then pursue a "logic" that seems designed to circumvent inconvenient and unwanted facts.

> Whereas mutations are usually deleterious to individuals, they are necessary to the continued survival of the population. That is, the environment that the population is adapted to at one time may not be the same at another. Mutation within a population represents a potential to meet new conditions as they arise. Put another way, mutations are often not fit for the environments they originated in, but they might provide the needed variation to survive in a new environment.[10]

Thus, we are informed that mutations can be expected to be deadly but that good ones do arise with a probability equal to that of a random change of letters managing to improve a Shakespearean sonnet. The mutated individual and/or its possible offspring will quickly die out -- but if the environment changes in some fortuitous manner at the same time, the offspring might gain a foothold.

Despite the compounding improbabilities of any kind of success, Stein and Rowe write that "mutation is the ultimate source of all variation within the gene pool"[11] and is the raw material for the operation of the evolutionary process.

Population Genetics

As the genes in a population are passed to the next generation in the sex cells, a new gene pool develops. If the original population is large, the next generation can be expected to be fairly representative of the population from which it sprang. But if the progenitor sampling is only a small part of the total population, then a representative mix of genes might not ensue in the offspring. Gene frequencies might change as a result of the statistical error associated with small samples. Such a deviation in gene frequency is called *genetic drift*.

A similar effect results from the *founder principle*, which involves a product similar to that of the sampling error of genetic drift. This principle applies when a small segment of a population leaves the original population to establish itself elsewhere. If the segment is not genetically representative of the original group, it will give rise to a population reflecting characteristics associated with the founders' gene frequencies. Survivors of disasters, as well as colonizers, might exhibit the founder principle.

The gene pool in humans also can be affected by culture since our culture sometimes decides for us who is an appropriate mate and who is not. Likewise, people have their own individual preferences (perhaps conditioned by culture). These factors can serve to increase the frequency and expression of recessive genes that otherwise might fail to be manifested. Random matings would preclude such a result, but truly random matings occur neither among humans nor anywhere else in nature.

Contributions to the gene pool also vary because of differential fertility rates; not all couples produce the same number of offspring. Associated with fertility rates is mortality. Mortality can occur at any stage of development. According to one estimate that took fertility, mortality and other factors into account, about one-half of all human zygotes (fertilized ova) never reproduce: Fifteen percent are lost before birth; 3% are stillborn; 2% are lost during the neonatal period; 3% die before maturity; 20% never marry (although some of these, admittedly, do bear or sire children); and 10% marry but remain childless.[12]

In short, natural populations are not in *genetic equilibrium*. Cited above are four mechanisms that supposedly bring about disequilibrium (changes in gene frequency): mutation; sampling error (genetic drift and the founder principle); nonrandom mating; and differential fertility (also called natural selection). These factors, working together, are viewed as the mechanism of evolution, and the shifting of gene frequencies is called evolution *even though the shift could be back and forth between two varieties of a characteristic within a species*. Now we see why "evolution" is both a confusing and a deceptive term, at least insofar as the general public is concerned. This "shifting" aspect of gene frequencies is actually observable but it has absolutely nothing to do with the creation of new species.

Stein and Rowe, however, believe that it does, for they write: "Any factor which brings about differences in fertility or mortality is a *selective agent*. A selective agent places *selective pressure* upon certain individuals within a population which results in a change in the frequency of alleles in the next generation. This is evolution."[13]

Speciation by Natural Selection

The Neo-Darwinists get themselves involved in some logical absurdities that reflect unfavorably, perhaps, on the neglect of logic in modern education, for if the reasonably well-informed American were as logic-literate as computer-literate, he or she would have no trouble detecting the fallacies.

The Neo-Darwinists say that individuals or populations having greater fertility and lower mortality rates than other individuals or populations in a shared habitat show a better fitness to that environment.[14] In turn, fitness is supposed to be a measure of how well the organism is adapted to the conditions imposed by the habitat. But since survival is determined by fitness and fitness is judged by nothing more than survivability, the Neo-Darwinist explanation clearly is an instance of the *argumentum in circulo*, or circular reasoning.[15]

American writer Edgar Allan Poe called circular reasoning "that idol which . . . demands from its votaries that they reciprocate cause and effect -- reason in a circle -- lift themselves from the ground by pulling up their pantaloons -- and carry themselves on their own heads, in hand-baskets, from Beersheba to Dan."[16] The evolutionary gradualists have been lifting themselves by their pantaloons and carrying themselves on their own heads; and yet, although many scientists have recognized the circular reasoning of Neo-Darwinism, the contorted airborne posture of the Neo-Darwinist (i.e., the illogic of his stated positions) seems to be hidden from most of our institutions of higher learning and from textbook authors.

Fitness aside, a biological mechanism for change in response to environmental pressures does exist. Conventional wisdom, going against the grain of logic as well as observation, holds that the possible extent of such change is virtually unlimited.

As noted, the development of differences between (or among) subgroups within a population might result from a number of factors: genetic drift and the founder principle; non-random matings; and, so they say, mutations. Theory further holds that when these groups occupy different niches or environments, "selective pressures" will be different and that, should interbreeding among groups be sharply reduced, *subspecies* or *races* would arise. (It follows that for a given species, such as *Homo sapiens*, there would be a single "racial norm" and that all such differences would be variations around that norm. Consequently, there would be, strictly speaking, no such thing as "race." And this is the attitude of many, and possibly most, anthropologists.)

Subspecies or races remain interfertile. But we are told that this status can change if certain barriers arise. There are geographical barriers, such as oceans, deserts, mountains, etc. There may be habitat isolation in the same area, such as that between tree-dwellers and ground-dwellers. Other factors may be behavioral isolation and the suggestion, at least, of mechanical isolation because of incompatibilities in the structures of the male and female sex organs. Various types of mortality, before and after birth, also come into play.[17]

So if intergroup matings are cut off, each subspecies may evolve separately along divergent paths, according to Stein and Rowe: "If the differences in gene frequencies become great enough, they will no longer be capable of successful reproduction with each other if the barrier is later removed. Thus the two groups may evolve into distinct subspecies and then into two distinct species."[18]

Here, then, strung together through a concatenation of "ifs" and "mays" (imaginary possibilities), is the so-called "grand unifying theory" of biology: gradual speciation by means of mutation and natural selection operating through countless eons of time, i.e., Neo-Darwinism. One could hardly be accused of levity if one confessed that such Neo-Darwinian flights bring to mind the "plausible story" method of inventing adaptive purposes for an organism's attributes (see Chapter I).

The Myth of Speciation by Natural Selection

Most species seem to be capable of extreme variation in the structure and general appearance (morphology) of individual members. But in all the years of observation in the field and experiment in the laboratory, not a single new species has been seen to arise as a result of selection. Artificial selection (breeding) has produced an amazing diversity of dogs -- but they have always remained dogs; any two breeds are capable of producing fertile offspring. Countless mutations have been introduced into *Drosophilia* (fruit flies), but the organisms always remain fruit flies.

Reporting on two *Drosophilia* experiments by Ernst Mayr, Jeremy Rifkin writes:

> Even with this tremendous speedup of mutations, scientists have never been able to come up with anything other than another fruit fly. More important, what all these experiments demonstrate is that the fruit fly can vary within certain upper and lower limits but will never go beyond them. In [Mayr's] first experiment, the fly was selected for a decrease in bristles and, in the second experiment, for an increase in bristles. Starting with a parent stock averaging 36 bristles, it was possible after thirty generations to lower the average to 25 bristles, "but then the line became sterile and died out." In the second experiment, the average number of bristles was increased from 36 to 56; then sterility set in. Mayr concluded with the following observation: "Obviously any drastic improvement under selection must seriously deplete the store of genetic variability . . . the most frequent correlated response of one-sided selection is a drop in general fitness. This plagues virtually every breeding experiment."[19]

The only objective evidence of speciation in the past is the fossil record. But some very unobjective claims have been made about the fossil sequence. Niles Eldredge and Stephen Jay Gould quote Moore, Lalicker and Fischer to the effect that "many lineages among fossils or various groups have been firmly established. These demonstrate documentary evidence of gradual evolution."[20] But to suggest that the fossils offer any such demonstration is only wishful thinking.

Dr. Pierre P. Grassé, respected as one of the world's greatest living biologists, suggests adding an epitaph to "every book on evolution":

> From the almost total absence of fossil evidence relative to the origin of phyla, it follows that any explanation of the mechanism in the creative evolution of the fundamental structural plans [displayed in a given phylum] is heavily burdened with hypotheses. . . . [T]he lack of direct evidence leads to the formulation of pure conjectures as to the genesis of phyla; we do not even have a basis to determine the extent to which these opinions are correct.[21]

Fossils of the horse offer the most familiar "proof" of Neo-Darwinism to be found in the geological record. But this proof proves to be an illusion. Rifkin quotes remarks made by MacBeth on a 1981 PBS-TV segment:

> About 1905 an exhibit was set up [in the American Museum of Natural History] showing all these horses. . . They were arranged in order of size. Everybody interpreted them as a genealogical series; there is no [evolutionary] descent among them. They were found at different times, in different places[,] and they're merely arranged according to size. But it is impossible to get them out of the textbooks. . . .
>
> As a matter of fact, many of the biologists themselves forget what they are. I had a radio debate with a paleontologist some years ago and when I said there were no phylogenies [examples of evolutionary descent] he told me I should go to the Museum and look at the series of horses. I said, "But professor, they are not a

family tree; they are just a collection of sizes." He said, "I forgot that."[22]

Encyclopaedia Britannica provides examples of "observable evolution." One is the divergent frequencies of the sickle gene among black populations of West Africa and the United States; environmental pressures responsible for the frequencies are different in the two areas. Another is the survival of severe weather by sparrows "not too big" and "not too little," as compared to the non-survival of "less fit" individuals.[23] Are these phenomena meant to imply the process of speciation? In the search for officially recognized examples of the biological route from amphibian to mammal, one is presented with inanities.

Is it by deception or by blindness that this evolutionary scheme is perpetuated? Why can its adherents not recognize that they are unable to offer a single shred of observational evidence in its defense; or see that, to the contrary, there is ample evidence to refute it? In any discipline not hamstrung by philosophical constraints, such myths as the Neo-Darwinists recite would have been laid to rest decades ago.

Pierre Grassé implies that survival of the gradualist evolutionary scheme can be attributed to *both* deception and blindness. "Through the use and abuse of hidden postulates, of bold, often ill-founded extrapolations," Grassé writes, "a pseudoscience has been created. It is taking root in the very heart of biology and is leading astray many biochemists and biologists, who sincerely believe that the accuracy of fundamental concepts has been demonstrated, which is not the case." [24] And he adds: "The explanatory doctrines of biological evolution [natural selection] do not stand up to an objective, in-depth criticism. They prove to be either in conflict with reality or else incapable of solving the major problems involved."[25]

Facts Relevant to the Concepts of Species and Speciation

Many scientists seem unwilling or unable to discriminate fact from hypothesis. There are no facts upon which to base the idea of speciation by natural selection. This supposed process does not even qualify as a theory. To conform with the spirit of scientific method, a theory should be based on at least one observation. No such observation has occurred in the case of Neo-Darwinian evolution, and speciation by natural selection has never been in position to qualify as more than a hypothesis, the step preliminary to theory formalization.

What does the observational evidence in fact suggest? There are two broad categories of evidence: that from the fossil record and that from living organisms.

With regard to the record in the rocks, the identification of species is shaky at best. *Paleospecies* (fossilized or other remains) are identified solely on the basis of morphology. The precariousness of morphological identification can be appreciated when one considers such physically similar living organisms as horses and zebras (different species) and the extreme divergence in morphology of Great Danes and Chihuahuas (same species).

As for the living, members of a sexual species can be identified by their capability of interbreeding and producing fertile offspring. Sometimes two different species can cross, but their offspring are either infertile or cannot breed true. They cannot, with minor exceptions, perpetuate themselves, and the line will die out.

So much for preliminaries. Several kinds of evidence and data interpretation deserve a close look.

I. Mutations.

It is admitted that most mutations are harmful. Observed single mutations appear to be neutral at best. Moreover, the supposed appearance of a neutral mutation in nature might well be the expression of a recessive trait (gene) not previously detected. A beneficial cumulative effect of individual beneficial or neutral mutations has never been observed, and experiments (exhibiting long-run harmful side effects) indicate that it is not to be expected.

II. Variation in Living Species.

Living species exhibit great variability, but none has ever been seen to change by selection into a different species. Adherents of the Synthetic Theory say that such changes take a long, long time and could not be identified in any given generation. Time has been reduced artificially, however, in fruit fly experiments and domestic breeding. Still, species persist in doggedly maintaining their identity. Induced mutations and selective breeding never produce a species different from its progenitors, whether those progenitors are immediate or many generations in the past.

III. Chromosomes and Living Species.

Most living species, as noted heretofore, are characterized by their own chromosomal counts and structure. A seemingly small difference in structure can be responsible for vast differences in life-forms. A structural difference in just one chromosome, together with a few genetic variances, differentiates humans and chimps at the chromosomal level.[26]

A change at the chromosomal level would underlie, *in most instances*, a change in species. And no such change could be random; random chromosome changes are as deleterious as random gene changes. Moreover, the change would have to occur at a fixed point in time (chromosome counts obviously are discrete), a fact that surely negates any idea of speciation through gradualism (speciation over time).

IV. The Fossil Record.

Nowhere in the fossil record is there to be found an orderly grading of one species into another. The most widespread claim of geologists is that the fossil record, not the evolutionary concept, is at fault, the intermediate fossil forms (long-sought "missing links") having been destroyed through geomorphic upheaval, erosion, and continued redeposition, as well as by differential preservation of representative types (i.e., some biological materials or traces are more readily preserved than others).

But whether or not any evidence for gradualism ever existed, it does not exist now. Whenever changes are exhibited in the fossil record, they are discrete. No objective support for speciation by natural selection can be garnered. And as to the supposed corruption of the fossil record, some Neo-Darwinists are taking a different view.

Eldredge and Gould profess to have rejected gradualism in favor of what they refer to as "punctuated equilibria." They accept the breaks in the fossil record as being real, reflecting the way evolution actually is recorded in the rocks (the breaks are not merely artifacts of a damaged crust). Eldredge and Gould claim that the record is much better than tradition dictates. They paint a picture of long periods of biological stability punctuated by "rapid" periods of speciation in isolated subpopulations. Speciation occurs so quickly that no one should expect to find it depicted by the fossil remains. Thus, nature creates the illusion of evolutionary jumps but in reality leaves an undisturbed chronology.

However, Eldredge and Gould go on to make a revealing confession: *Both gradualism and punctuated equilibria are interpretations based on preconceived ideas.*[27]

To put the case precisely: What Eldredge and Gould did was to contrive an *ad hoc* hypothesis, "punctuated equilibria," to explain gaps in fossil evidence while still working within the governing hypothesis of speciation by natural selection. Punctuated equilibria would be an acceptable proposal only if there were evidence to support the governing hypothesis -- and there is not.

By "running gradualism at fast forward," Eldredge and Gould are able to characterize the "geological camera" as being unable to record the tracks of evolution. They are able to have their cake and eat it, too: The geological record is not so bad after all, and the dogma of speciation by natural selection is protected even if not demonstrated.

The work of Eldredge and Gould (and their fellow Neo-Darwinians) has all the air of that project, described by Jonathan Swift in *Gulliver's Travels*, for "extracting sunbeams out of cucumbers, which were to be put into vials hermetically sealed, and let out to warm the air in raw inclement summers."[28] Sunbeams can be extracted from cucumbers as easily as gradualism from the geological record, with about as much promise of shedding any light.

There is no continuum of fossil evidence. Thus, for the Neo-Darwinian, anything that can serve as a transitional form becomes of paramount importance. The misrepresentation of the horse in this regard has already been cited. To the horse can be added the well-known "reptile-bird," *Archaeopteryx*. This creature had characteristics that appear reptilian -- teeth, small claws on its wings, and a long vertebrate tail -- but it also seems to have had feathers. It is routinely hailed as a transition between reptiles and birds; it is adjudged to have been a poor flyer, perhaps only a glider.[29]

But in 1977 a fossil bird was found in rock strata from the same geological period as *Archaeopteryx*.[30] The "ancestor" of birds turned out to be a cousin. Furthermore, claw-winged birds are not merely extinct relics of the distant past. Along the Amazon River today is found a bird called hoatzin, a clumsy flyer that "takes wing with great difficulty and often crash-lands a few hundred feet away." The account continues:

> The baby hoatzin is born with two claws on each wing and

> uses them as forelimbs to climb trees with agility within a few hours of birth. If it is threatened, the bird leaps from its nest into the river and swims to cover. When the danger has passed, the baby hoatzin pulls itself back to its nest with its clawed wings. Although the claws drop off as the bird approaches maturity, fully grown hoatzins continue to use their wings for climbing through foliage.[31]

As is the hoatzin, *Archaeopteryx* was a "finished product," not some imagined Darwinian linkage. (This does not mean, however, that *Archaeopteryx* could not have been an ancestral form. This is addressed in Chapter VI.)

It would be a mistake, however, to debunk the fossil record entirely, as many creationists are prone to do. Despite faulty assumptions about the fossil record and many erroneous interpretations of it, the fossil record does contain a wealth of positive information. It shows that new forms of life have arisen at various times in the past. At one epoch after complex life-forms proliferated, there remained, as yet, no birds. Once there were no humans, albeit many large forms of life roamed the earth. The record also shows that once-extant life-forms, such as *Archaeopteryx*, are no more. Moreover, changes have occurred on a grand and global scale. But the evidence for such changes does not conform to a uniformitarian framework.

Conservation of Species

A species strives to preserve itself. More accurately, perhaps, one might say that it is dynamically programmed to facilitate its survival; it "flexes" with its environment. Rifkin, quoting Grassé, writes:

> . . [M]utations are "merely hereditary fluctuations around a medium position: a swing to the right, a swing to the left, but no final evolutionary effect . . . they modify what pre-exists." Whereas Darwin thought that variations led to new species, the evidence proves the contrary: namely, that *variation improves the ability of the species to maintain itself against radical change.*

Grassé concludes: "Once one has noticed microvariations (on the one hand) and specific stability (on the other), it seems very difficult to conclude that the former (microvariations) comes into play in the evolutionary process." Grassé says that the evidence forces us "to deny any evolutionary value whatever to the mutations we observe in the existing fauna and flora."[32]

This "fluctuation" or "microvariation" is what science has come to view as the operation of natural selection. But this "flexing" stops far short of speciation -- and probably accomplishes its adaptive work without the necessity of genetic mutations. Random mutation, as observed, is generally harmful. The significance of those mutations that are not deleterious is uncertain. It could be that much of the potential for this type of change emerges along with the species itself.

To play the "may," "might," "could be" game of the Neo-Darwinists: There are possibly no significant genetic alterations (baneful, neutral or beneficial) outside the normal replication and reproduction processes. Rather, as far as anyone has been able to determine through observation, species manifest changes in gene frequencies at the population level. This genetic drift typically takes place over multiple generations and produces phenomena such as that observed in the peppered moth.

> [The peppered moth is] a European moth . . . having speckled black and white wings. It is of significance in exemplifying natural selection through industrial melanism. A dark (melanic) form of the peppered moth, first noticed in Manchester, England, in 1848, by 1898 had outnumbered the usual light-colored moth by 99 to 1. The explanation of this phenomenon is that the dark moth was not as conspicuous to bird predators as the light moth against the black-sooted tree trunks of industrial areas.[33]

The species thus maintains its identity by means of natural selection. MacBeth provides more on this topic, quoting other authorities:

> ". . . [N]atural selection . . . is . . . usually and most strongly a stabilizing, normalizing influence preventing or slowing and not hastening evolutionary change." The same view is expressed by Williams ". . . I regard it as unfortunate that the theory of natural selection was first

developed as an explanation for evolutionary change. It is much more important as an explanation for the maintenance of adaptation . . . evolution takes place, not so much because of natural selection, but to a large degree in spite of it ." [34]

Certain biological norms thus can alternate, flourish at the same time, or even alternate or co-exist with additional norms (as in dog breeding). Two different colors of moth, for example, can flourish simultaneously in their respective (different) ecological niches. Or one variety might disappear entirely, always with the potential of re-emerging. Moreover, this process might be driven to certain morphological extremes, notably in the case of artificial selection.

But always, within a species, all varieties remain interfertile unless, as noted in the case of fruit flies, a line is pushed too far resulting in an inability to reproduce. If all the varieties within a species were thrown together and completely random mating were achieved, the species could be expected, after many generations, to revert to the original ancestral type. And this type could be used (*ideally*) to replicate the same morphological variety that existed before all the varieties were thrown together and intermixed.

These are, of course, ideal results. Completely random mating is not seen in nature, and throwing all of the varieties together might well result in a half-dozen basic types, based on selective mating, rather than in the ancestral original. The point, however, is that no creature on Earth, including man, is "evolving" into something different. Arguments to the contrary are unfounded.

Conclusion: Conjuring with a Circle

The reader impressed by the dismantling of the Neo-Darwinist system might well ask: "Why this never-ending debate? If Darwinism is as faulty as many claim, can the fallacies not be demonstrated to the satisfaction of all?"

The answer is "No, they cannot." No event, observation, argument or organism in nature can prove the theory of natural selection to be false; Neo-Darwinist theory forms a perfect, self-supporting circle. It is a closed system -- an impenetrable bubble floating in air. The system says that if any organism survives and leaves offspring, it is "fit," and because it is fit, it survived. Indeed, Neo-Darwinist theory is a perfect example of what Samuel Taylor Coleridge called a "comprisal of the principium principii in the argumentum in circulo." The Neo-Darwinist begins by begging the question and ends by going in a circle back to the unfounded assumptions with which he starts.

The Neo-Darwinist's "comprisal of the principium principii in the argumentum in circulo" is not shielded by the kind of usefulness that invests, for example, Newton's laws of motion. The difference between Newton's method and that of Darwin is in the "principium principii," i.e., the question-begging.[35] By avoiding the attempt to explain gravitation and contenting himself with discovering "how to measure [the] inertial and gravitational properties [of matter] in terms of acceleration, which means in terms of distance and time,"[36] Newton devised tools for prediction that, according to Werner Heisenberg, "are an essential part of the language which forms the basis of all natural science."[37] By contrast, Darwinism describes only an inferential construct from the fossil record, not observable ongoing reality.*

Neo-Darwinism is a religion conjuring with a magic circle. The theory of natural selection cannot be refuted in principle, for it is a semantic disguise for a secular faith. This fact is well-illustrated in Tom Bethell's remarks on the verification of Darwinism by means of observations on the peppered moth. This moth is cited by *Encyclopaedia Britannica* as a "striking example" of evolutionary change by natural selection.[38] Tom Bethell disagrees:

> . . . [T]his observed relationship between moth-color and moth-survival does not constitute a "verification" of Darwin's theory that the fittest survive by virtue of some independent isolable quality of "fitness." It merely verifies the theory that camouflage is helpful to moths in a particular environment. One can easily appreciate this point by considering the biologist's reaction if the camouflage hypothesis were, much to his surprise, falsified; that is, if dark moths in another part of the woods (or the world) did not survive as abundantly as other, more visible strains even though the tree trunks remained dark and other aspects of the environment

* See "A Note on Predictive Power" at the end of this chapter.

> seemed similar. Would the biologist, faced with such a finding, triumphantly report that he had falsified Darwin's theory of natural selection? No. He would merely conclude that some other factor he did not know about was operating in the environment -- one that he had not hitherto considered.
>
> It follows, then, that darkness in moths is not an independent criterion of fitness, leading ineluctably to survival and offspring, but merely one factor (among thousands) that contributes to survival in a given environment, or set of circumstances.[39]

In short, Darwinism -- i.e., speciation by selection (whether natural or otherwise) -- is a hypothetical process, never observed; it always has been, and remains, an article of faith. It is science's equivalent of science's anathema: religious fundamentalism. It is not science, and the credibility it is accorded is undeserved. The same is true for uniformitarianism in general, the philosophical edifice housing Darwinism.

The inadequacy of Darwinism perhaps is finally coming into consciousness, else there most likely would be no discussion of "punctuated equilibria." But unless the revisions in New Synthesis thought can accommodate, as an explanation of the fossil record, the sudden emergence of new and completely integrated life-forms instead of an accelerated gradualism, recent evolutionary speculation will prove fruitless.

A NOTE ON PREDICTIVE POWER

For those who do not know the history of science, the difference between Newtonian mechanics and Darwinian metaphysics may not be obvious.

Immediately after Newton published *Philosophiae Naturalis Principia Mathamatica*, in which he propounded his ideas on gravity, Leibniz accused him of having written a book that "deserts Mechanical causes, is built upon Miracles, & recurrs to Occult qualitys."[40] Leibniz wanted causes; Newton offered only description, a quantitative formulation of how the force of gravity acts, disavowing any intention of *explaining* that force. Without explanation, the force would have to remain an "occult quality," as Leibniz characterized it.

On the other hand, the "miracles" to which Leibniz referred began to vanish rapidly in the work of Lagrange and Laplace. Newton himself had felt that the intercession of God was necessary to "maintain the stability of the solar system in the face of perturbations of planetary motions by other planets."[41] But Joseph Louis Lagrange, using Newton's laws, mathematically demonstrated that observed irregularities in the motions of Earth and its moon and of Jupiter and its moons "were an effect of gravitation."[42]

Pierre Simon Laplace then proved "that the irregularities in the eccentricities of the elliptical paths of the planets were periodic. That is, these irregularities would oscillate about fixed values and not become larger and larger and so disrupt the orderly motions of the heavens."[43]

The predictive power of Newtonian mechanics was given a signal demonstration by two astronomers, John Couch Adams and U. J. J. Leverrier. Using the astronomical theory of Lagrange and Laplace, Adams and Leverrier predicted the existence and location of the planet Neptune by deducing "the mass and path of an unknown [never observed and therefore theoretical] planet from its effects on the motion of Uranus." Neptune "was barely observable with the telescopes of those days and would hardly have been noticed if astronomers had not been looking for it at the predicted location [and time]."[44]

The success of Newtonian mechanics should be compared with the utter lack of success by Neo-Darwinian evolutionism. Darwinesque gradualism has never predicted anything; this "missing link" in performance should be ringing alarm bells in a lot of Wise Men's heads, especially in view of the predictive power of Velikovsky's theories.[45]

REFERENCES

1. Richard Dawkins, "The Necessity of Darwinism," *New Scientist* (15 April 1982), p. 130.
2. *Encyclopaedia Britannica*, 15th ed., "Evolution," Vol. 7, p. 13 (emphasis added).
3. Niles Eldredge, "Gentleman's Agreement," *The Sciences* (April 1981), p. 20.
4. Norman MacBeth, *Darwinism: A Time for Funerals*, Robert Briggs, Asc. (San Francisco 1982), pp. 2-3.
5. Garland Allen, *Life Science in the Twentieth Century*, Cambridge University Press (Cambridge 1978).
6. *Encyclopaedia Britannica*, 15th ed.
7. Alexander Rosenberg, *The Structure of Biological Science*, Cambridge University Press (Cambridge 1985).
8. Arthur O. Lovejoy, *The Great Chain of Being*, Harvard University Press (Cambridge 1936, 1964), passim, especially Chap. VIII, "The Chain of Being and Some Aspects of Eighteenth-Century Biology."
9. P. L. Stein and B. M. Rowe, *Physical Anthropology*, McGraw-Hill (New York 1982), pp. 80-1.
10. Ibid.
11. Ibid., p. 81.
12. Ibid., p. 90.
13. Ibid., p. 95.
14. Ibid.
15. Norman MacBeth, "How to Defuse a Feud," *KRONOS* VII-4 (Summer 1982), pp. 1-2.
16. Edgar Allan Poe, *Marginalia*, University Press of Virginia (Charlottesville 1981), p. 151.
17. Stein and Rowe, op. cit., pp. 107-8.
18. Ibid., p. 106.
19. Jeremy Rifkin, *Algeny*, Viking (New York 1983), p. 134.

20. Niles Eldredge and Stephen J. Gould, "Punctuated Equilibria: An Alternative to Phyletic Gradualism," *Models in Paleobiology*, ed. by T. J. M. Schopf (San Francisco 1972), p. 91.

21. Pierre Grassé, *Evolution of Living Organisms* (New York 1977), p.31.

22. Rifkin, op. cit., p. 129.

23. *Encyclopaedia Britannica*, 15th ed., Vol. 7., p. 15.

24. Grassé, op. cit., p. 6.

25. Ibid., p. 202.

26. Stein and Rowe, op. cit., pp. 295-6.

27. Eldredge and Gould, op. cit., pp. 96-8.

28. Jonathan Swift, *Gulliver's Travels*, in *Gulliver's Travels and Other Writings by Jonathan Swift*, ed. by Louis A. Landa, Riverside Press (Cambridge 1960), p. 145.

29. *Encyclopaedia Britannica*, 15th ed., "*Archaeopteryx*," Vol. 2, p. 1059.

30. "Bone Bonanza: Early Bird and Mastadon," *Science News*, 112 (Sept. 24, 1977), p. 198.

31. *Rivers and Lakes*, by the editors of Time-Life Books (Chicago 1985), p.105.

32. Rifkin, op. cit., pp. 132-3.

33. *Encyclopaedia Britannica*, 15th ed., "peppered moth," Index VII, p.863.

34. Norman MacBeth, *Darwin Retried: An Appeal to Reason*, Gambit (Boston 1971), pp. 48-9.

35. Morris Kline, *Mathematics in Western Culture*, Oxford University Press (London 1953), p.211.

36. Unlike Newton, Darwin did not disclaim hypotheses. In *The Autobiography of Charles Darwin*, ed by Nora Barlow, Norton (New York 1958), Darwin wrote: "Some of my critics have said, 'Oh, he is a good observer, but he has no power of reasoning.' I do not think that this can be true, for the *Origin of Species* is one long argument from the beginning to the end, and it has convinced not a few able men" (p. 140). Unfortunately, Darwin rejected the catastrophic hypotheses favored by his former mentor, Sedgwick, in favor of natural selection over vast stretches of time.

37. *The Encyclopedia of Philosophy*, "Newtonian Mechanics and Mechanical Explanation," Vols. 5 & 6, Macmillan (New York 1967), p. 496.

38. *Encyclopaedia Britannica*, "peppered moth," op. cit.

39. Tom Bethell, "Darwin's Unfalsifiable Theory," *KRONOS* VII-4 (Summer 1982), pp. 34-6.

40. I. B. Cohen, "Foreword" to Isaac Newton, *Opticks*, Dover (New York 1952), p. xxix.

41. *The Encyclopedia of Philosophy*, "Newton, Isaac," Vols. 5 & 6, p. 490.

42. Kline, p. 209.

43. Ibid., p. 210.

44. Kline, p. 210.

45. See Thomas Ferté, "A Record of Success," *Pensée* II-2 (May 1972), pp. 11-15, 23.

CHAPTER IV

Noah's Flood: A Problem for Both Camps

Ancient records from around the world tell of a near-total destruction of mankind in the distant past by a worldwide deluge. Also preserved in many of the stories are memories of a pre-Deluge *Golden Age*, a time when much of the world was like a garden. Climates, seasons and weather had little variability. Man was one with nature and lived on the fruit of the land with the expenditure of little or no effort. Then the Flood came. Conditions became harsh. The antediluvian world became mankind's paradise lost.

The most familiar account of such a disaster is of course that in the Book of Genesis. There we read that Noah, who alone "finds grace in the eyes of the Lord," is forewarned of the coming flood and told to build a huge ark or, more correctly, chest (Heb.: *tebah*) in which to escape with his family, taking a representation of all animal life with him (Gen. 6:8-22).

The uniformitarians deny that such a catastrophe ever occurred, demanding, "Where is the evidence? Where could the water have come from?" But no evidence is allowed to contradict uniformitarian assumptions.

The Flood is central to creationist thought; it is believed by most creationists to have been the only global cataclysm in the whole history of Earth. Henry Morris writes that "all fossiliferous deposits have been formed sometime within the space of human history. There seems, therefore, no better explanation for their existence in most cases than the Flood and its associated geological and hydraulic activities."[1] Says S. E. Nevins, "[T]he Biblical account offers the simplest proposal -- a single creation and a single catastrophe."[2] According to F. L. Marsh, "[T]he fossil record is not a historical record as the evolutionist philosophers would portray it. It but reveals to us fragments of some of the forms which were living concurrently at a time about seventeen centuries after creation."[3]

R. L. Wysong sums up the opposed positions: "The classical geological column and timetable spread the record of the fossils over millions of years. The creationist looks at the same data and concludes that the 'geological timetable' is a record of one large devastating flood that accomplished the majority of its work in a few months. . ."[4]

Assuming that the widespread testimony of many ancient peoples lends some credence to the story of the Flood, one still must question the position of the creationists. It is true that the Bible explicitly describes only one catastrophe, the Deluge, as global in scope, but why must one necessarily assume that the Bible allows or implies no others? For example, if "the sun stood still" at Joshua's command, as alleged in the Book of Joshua (10:12-13), would not the whole world have been affected? [5]

Does the biblical story of the Flood give the complete picture? What light might *other* records throw upon this cataclysm?

Non-biblical flood legends contain many similarities to each other -- and just as many differences. Some of the stories seem to be so incoherent or fairy-tale-like in nature that the most open-minded analyst might be driven to question their value as historical evidence. Was there really a global deluge in antiquity?

Various elements of the biblical account appear in the nonbiblical versions, but there is little or no consistency in their assemblage; one element might appear in one account, a different element in a second account, and perhaps both of these same elements in a third. A direct categorization of the legends that might result in any significant insight does not seem to be possible; however, a simple dual classification followed by a statistical treatment is most revealing. It is ironic that these legends prove not only to support the essentials of the biblical record, but also to challenge the creationist viewpoint in one or more respects.

Statistical Analysis

The analysis that follows is based upon a collection of sixty-one (61) legends from all quarters of the world. These were first classified according to whether they speak of a favored family, group or individual (cf. Noah) being saved from a flood or else tell of some unspecified remnant somehow surviving (this latter is not always an obvious choice because of the form in which some of the legends or myths were couched). With the biblical version as a model -- and excluding it from the data -- various elements were then examined for correlation with the Family/Remnant classifications.

Five characteristics were considered. They are summarized below and tabulated for comparison, then analyzed for statistical results.

I. Summary of the Data.

In thirty-six (36) stories, a favored family, group or individual is saved. In twenty-five (25) stories, an unspecified remnant survives.

Of the thirty-six involving a favored family, group or individual:

Eighteen describe survival in a boat.

Thirteen describe other means of survival.

Nine indicate a preservation of other seeds of life.

Fifteen indicate being saved by a specific forewarning.

Twelve describe other disturbances besides the flood.*

Twelve indicate that the flood came by divine decree.

Of the twenty-five stories involving an unspecified remnant:

Six describe survival in a boat.

Eight describe other means of survival.

Two indicate a preservation of other seeds of life.

Ten describe other disturbances besides the flood.

Twelve indicate that the flood came by divine decree.

PERCENT OF TOTAL CATEGORY

	Means of Survival	Other Life	Forewarning	Other Disturbances	Divine Decree
Family, etc.	50%/36%	25%	42%	33%	33%
Remnant	24%/32%	8%	8%	40%	48%

II. Summary of Results.

The analytical procedure initially assumed no correlation between the selected characteristics and the category of Favored Family, etc. The statistical test then attempted to reject this assumption. As in any statistical analysis, the outcome must be qualified by a statement of the probability of error. The elements being evaluated against the category of Favored Family were those listed above, to wit: means of survival, preservation of other life-forms, specific forewarning, other natural upheavals associated with the flood,* and "flood decreed by God." The results of the test were as follows:

TABULATION OF RESULTS

Assumption: There is no contingency upon the Favored Family by the elements:	Verdict ("Reject" means to reject the idea of independence -- i.e., reject "no contingency")	Probability that rejection is the wrong decision.
Survival by boat	Reject	2.5%
Forewarning	Reject	0.5%
Other life saved	Reject	10.0%
Other upheavals	Do not reject	--
Decreed by god(s)	Do not reject	--

The only "failures" are in not rejecting the assumed independence between Favored Family and each of the last two elements, Other

* Earth movements, volcanic eruptions, celestial disorder, etc.

Upheavals and Flood Decreed by God(s). "Do not reject" does not mean "accept"; it means there is not enough statistical correlation to support a conclusion. There does appear to be a strong correlation between three of the five elements and the concept of a Favored Family.

Geographical distribution of the categories Family and Remnant was the next consideration that was explored. This tabulation came out as follows: Europe: seven cases of a family and two of a remnant; Middle East: five cases of a family and one of a remnant; Africa: One case of a family and two of a remnant; Asia: Three cases of a family and one of a remnant; Pacific Islands: Four cases of a family, three of a remnant; The Americas: Sixteen of a family, sixteen of a remnant. Tabulating the results reveals an interesting picture:

DISTRIBUTION OF CATEGORIES BY PERCENT OF TOTAL

	Europe	Middle East	Africa	Asia	Pacific Isles	The Americas	World-wide
Family	78%	83%	33%	75%	57%	50%	59%
Remnant	22%	17%	67%	25%	43%	50%	41%
Approx. Ratio	3.5	5.0	0.5	3.0	1.3	1.0	1.4

Conclusion: A Real Event

Since the characteristics checked for independence were taken from the biblical account -- which itself was not included in the data -- the results of the analysis suggest that the biblical version, where each association is explicit, was recorded at a very early date.

The geographical distribution of the memory of the Favored Family (Story of Noah) could have come about in at least two different ways. The first one that comes to mind is diffusion, which simply would entail "export" by the migrations and movements of those people (or their descendants) who had the experience.

But the same distribution can be explained equally well by an entirely different conception of "Noah and the Ark," as envisioned by Dwardu Cardona. Cardona has suggested that Noah and the Ark were a celestial apparition that would have been visible throughout the

Northern Hemisphere.[6] If such an image were sufficiently distinct, it could inspire many similar accounts in many different places at once. As far as our statistical analysis is concerned, the results would be the same. No discernment of the *means* of distribution is possible from the tables.

If Cardona's interpretation is correct, then the Noah-type stories should predominate and be concentrated in Earth's Northern Hemisphere, New World as well as Old. Other less specific or more fanciful descriptions of the apparition likewise should be from above the equator or from not far below it. (Observations of a shorter duration might have resulted from below the equator if Earth's axis temporarily tilted enough.) A simple tabulation of fifty-four (54) of the accounts might provide support for Cardona's view.

	Number of stories with favored individual(s) or an apparent celestial apparition.	Number of stories that are non-specific beyond an account of a flood.
Northern Hemisphere	40	9
Southern Hemisphere	1	4

Different experiences clearly are reflected in the diversity of aspects preserved in the many different accounts. These different experiences must have resulted from separations in space (geographic location), in time, or in both; the various legends collectively, and in some cases individually, embody memories from more than a single perspective. (Although not explicit in the Genesis account, the representation of more than one perspective also might be inferred from other parts of the Bible.) The evidence seems to indicate either that catastrophic flooding of global or near-global dimensions occurred more than once; that there were more survivors of the Great Deluge than a single crew (the second event is implied by Cardona's interpretation, in which the "crew" were not the actual survivors); or that both conditions are to be inferred.

If there were several catastrophic events, the various accounts perhaps indicate that, although many lives were lost in the later catastrophes, the many scattered survivors (all around the globe) of the far more ruinous Great Deluge undoubtedly retained traditional oral accounts of it; and, further, that these traditional accounts subsequently were mingled by later generations with orally preserved memories of the more recent upheavals.

The Flood has other negative ramifications for the creationist. One of these is revealed by a consideration of the extremely restricted world of the yucca moth:

> Each yucca moth species is adapted to a particular species of yucca [plant]. The moths emerge when the yucca flowers open. The female gathers pollen from one flower, rolls it into a ball, flies to another flower, lays four or five eggs, and inserts the pollen mass in the opening thus formed. The larvae eat about half the 200 seeds produced by the plant. The yucca can be fertilized by no other insect, and the moth can utilize no other plant.[7]

The creationists cite this biological interrelationship as a proof of creation. It is another refutation of Darwinism -- but if the result of creation, then creation when?

Given a universal destruction by the Flood, the relationship must have come into existence afterward; it would be absurd to maintain that the two partners migrated in unison from an ark in the Old World to their home in the New, surviving only in the latter. The alternative is something less than universal destruction.

To the uniformitarian: The many records left by our forebears on the nature of the flood(s) refute uniformitarianism.

To the creationist: The record in the earth's strata cannot be attributed to a single flooding catastrophe; the evidence argues against the claim of *one* flooding catastrophe *and* the destruction of all life.

The Darwinist ignores the true nature of the evidence, and the creationist contradicts himself in his interpretation of it.

The Great Deluge was a apparently real event, whatever its total scope and magnitude; and it was only one of several cataclysms that scarred the face of the world in ancient times. We look now to the earlier upheavals -- a time long before man walked upon the earth.

REFERENCES

1. H. M. Morris, *Studies in the Bible and Science*, Presbyterian and Reformed Publishing Co. (Philadelphia 1966), p. 134.

2. S. E. Nevins, "A Scriptural Groundwork for Historical Geology," from *A Symposium on Creation II*, Baker (Grand Rapids 1970), p. 100.

3. F. L. Marsh, *Life. Man. and Time*, Outdoor Pictures (Escondido 1967), p. 135.

4. R. L. Wysong, *Creation-Evolution: The Controversy*, Inquiry Press (East Lansing 1976), p. 365.

5. See discussion by Immanuel Velikovsky, *Worlds in Collision*, Doubleday (Garden City 1950), pp. 39-46.

6. Dwardu Cardona, "The Road to Saturn," Part I, *Aeon* I-l (January 1988), pp. 16-17.

7. *Encyclopaedia Britannica*, 15th ed., "Incurvariidae," Index V, p. 326.

BIBLIOGRAPHY FOR THE STATISTICAL ANALYSIS

S. N. Kramer, *History Begins at Sumer*, Doubleday (Garden City 1959).

R. W. Rogers, *Cuneiform Parallels to the Old Testament* (New York 1912).

Lucian (A.D. 120-180), *The Goddess of Syria*.

Larousse Encyclopedia of Mythology, 2nd ed. (London 1968).

Pindar (522-433 B.C.), *Olympian Odes*.

Apollodorus (c. 150 B.C.), *Library*.

J. P. Cohane, *The Key*, Crown Publishers (New York 1969).

Ovid (43 B.C.-A.D. 17), *Metamorphoses*.

Thorpe and Blackwell, *The Elder Eddas* (London 1906).

L. de Cambrey, *Lapland Legends* (New Haven 1926).

E. E. Clark, *Indian Legends of the Pacific Northwest*, University of California Press (Berkeley 1960).

Byron Nelson, *The Deluge Story in Stone*, Augsberg Publishing House (Minneapolis 1931).

PART TWO

Origins and Non-Origins: New Perspectives

INTRODUCTION TO PART II

The means of the development of life on Planet Earth is still an open question. The two most widely held points of view are not compatible with known relevant facts. There is no way these viewpoints can be expected to explain the facts.

We shall see that past catastrophes -- global upheavals -- appear to have embodied whatever trigger is necessary for biological change. The primal trigger -- the first cause -- is more difficult to conceptualize in a scientific context, if not impossible.

It should be obvious that the effects of universal disturbance would not be confined to biological systems. Moreover, man's interest in origins is not restricted to the field of biology. The accepted theories of origins in general are as impotent as Darwinism is in explicating the process of speciation. But this is not to say that the circumstances of all origins are undiscoverable. Sometimes it is a matter of *how* we look, as much as *where* we look.

In search of various origins, along with certain "nonorigins," we examine now a range of phenomena, remaining receptive to all available evidence and testimony.

CHAPTER V

The Signature of Catastrophe: The Origin and Non-Origin of the Geological Record

Our species has witnessed a number of violent disturbances of the planet on which we dwell. According to ample historical testimony, one such disturbance, the Great Deluge, was nearly the cause of our extinction. Uniformitarian critics offhandedly dismiss the idea of a worldwide flood on the grounds that Earth's atmosphere could not hold enough water for it. The creationists have found comfort in their vapor canopy, but this idea doesn't hold water either, so to speak.
Velikovsky looked beyond Earth's atmosphere for the source of the floodwaters. Relying on legends of the sort cited in the last chapter, as well as on analysis of observed facts (including the existence of hydrogen and water elsewhere in the solar system), Velikovsky concluded that "the Earth became enveloped in waters of cosmic origin, whether coming directly . . . or formed from clouds of hydrogen gas . . . which combined . . . with the Earth's own free oxygen."[1] He added that sea levels increased substantially and the Atlantic Ocean was created thereby (or, at least, attained a new significance).

The Great Flood was a drowning of the earth. Since then, there have been phenomenal electrocutions and fearful, thunderous

THE GEOLOGICAL COLUMN (TIME SCALE)

ERA	PERIOD	EPOCH	
Cenozoic	Quaternary	Holocene	
		Pleistocene	2
	Tertiary	Pliocene	5
		Miocene	23
		Oligocene	37
		Eocene	55
		Paleocene	65
Mesozoic	Cretaceous		140
	Jurassic		195
	Triassic		225
Paleozoic	Permian		280
	Pennsylvanian		320
	Mississippian		360
	Devonian		400
	Silurian		435
	Ordovician		500
	Cambrian		570
Precambrian Eon			

Figures represent millions of years before the present (conventional dating).

earthshocks and bombardments by extraterrestrial materials, as described in the books of Exodus and Joshua. [2] Specific instances are noted in later chapters.

As nature ran wild, geography, directions, seasons and even the bases of our reckoning of time were altered.[3] It is difficult to fathom the impressions that must have been burned into the minds of our ancestors. Who can conceive of the fear of the sky falling? The world exploding? The destruction of all Creation?

Creation was never totally destroyed, of course, and the onslaught finally came to an end. Things now have been tranquil for so long that the gradualist can deny that Nature ever went on a rampage -- or that the records left by our ancestors have any basis in reality.

Yet there is another record -- the record in the rocks. The earliest "account" in Earth's crust is older by far than the accounts in books.

A few words about planet Earth, the place that we call home, are in order. Clearly, it dominates every aspect of our lives, dwarfing the human creatures who populate its surface; in the end, it consumes even our most ambitious works. Yet, almost belying its frequent fits of violence, Earth stands as the sure, stable foundation of mankind's existence.

Although a man is as a gnat -- nay, a microbe-- on the surface of Earth, this massive dwelling place pales to insignificance in the context of the universe; its physical presence is far less in relation to the whole universe of matter than a single man is in relation to Earth. Even within our own solar system, Earth is distinguished by its unique biosphere, not its size. Earth's distant anchor, the incomprehensibly larger Sun, is lost in a sea of countless suns (many far larger than ours) making up the Milky Way Galaxy -- which, in turn, is lost in a maze of galaxies. The size and significance of man's world are thus matters of perspective.

So much available space unquestionably works to our advantage. We are separated from other planet-size bodies by hundreds of thousands to hundreds of millions of miles. We move around the sun in our own private, tranquil path, greeted by no more than an occasional shower of meteors; we have learned not to expect otherwise. To encounter another body the size of Earth would be a horror unimaginable.

Beneath our feet in the crust of the earth is another world, that of the ancient past. Locked in place, seemingly outside of time, are the lithified remains and traces of many creatures that no longer walk the world's surface or swim in its waters or fly in its heavens. They are packed in layer upon layer of sedimentary rock. There they remain until disturbed by forces of nature or, more recently, by the devices of mankind. How did these creatures come to such a fate, and when?

Orthodox science maintains that their burial and accumulation was a slow process spanning many millions of years, millions of generations ago. The party line says that many of these fossilized beds were deposited in shallow seas that once extended over large areas of the present-day continents and that those beds now above sea level testify to a subsequent vertical shift of Earth's crust. This depositing of the geological record supposedly took place at a rate undetectable by any living creature. The conditions necessary for preservation being rare, millions upon millions of years are marshaled by current theory to account for layers of remains measuring thousands of feet thick. The discipline of historical geology attempts to interpret the nature of such processes and events and to reconstruct their sequence in a model called The Geological Time Scale or, as preferred here, The Geological Column.

The Standard Geological Column

The Standard Geological Column (see figure) presents a very precise (not to be confused with accurate) picture of the geological record and of events recorded in the crust of the earth. The Column is a highly subjective representation, constructed piecemeal under uniformitarian assumptions. Its divisions seldom represent universal physical divisions in Earth's multitudinous layers of crustal rocks. Instead, the gradualists have turned logic on its head. Basing their classifications largely on the assumption of the continuous evolutionary change of species through time, they use fossils found in sedimentary rock as the primary means of correlating layers of rock and of establishing a standardized time scale for all locales of the earth.

In the geological record are two different kinds of unit. Rock-stratigraphic units, known as *geological systems*, are defined based on selected sequences of layered rock in type areas (a type area being the first representation of a particular sedimentary sequence that is discovered). The *geological period*, a term denoting time, is inferred from the existence of a geological system. The system itself might appear in only one place on Earth. Periods are grouped into larger time units called *eras*.

The earliest recognized era is the Precambrian, in which the evidence for life is inconclusive. The three later eras -- Paleozoic (ancient life), Mesozoic (middle life) and Cenozoic (modern life) -- bear fossils in abundance.

The boundaries of many systems are problematical because they have not been clearly stipulated or because type areas of "vertically adjacent" systems are widely separated geographically. In the initial definitions of systems, a search was made for natural boundaries. These boundaries frequently derive from drastic changes in either lithology (rock type) or fossil content, i.e., from a pronounced physical break. Boundaries are determined today largely on the strength of gradational changes in fossil content. Fossils also are used as boundary indicators when fossils appear in higher strata after never having shown up in lower ones or when fossils disappear in higher strata after having been prevalent in lower ones. Frequently, however, the level of a natural boundary in one area of the globe differs from that in some other area. The assignment of such uncertain strata to one system or another is arbitrary. Controversies continue. Sometimes, they can be resolved only by international agreement.

Since examples of "boundary situations" can be instructive, several of these follow.

Paleozoic Era. The beginning of the Cambrian Period is marked by the widespread appearance of marine invertebrates of many groups. Unable to settle on a boundary between the Cambrian and the Silurian, scientists in 1879 defined the Ordovician to embrace the overlap of the two. The Ordovician is characterized not by unique fossils, but by its fossil groupings. Its rocks have no singular characteristics. Similarly, the Devonian was proposed to resolve ambiguities between the Silurian and Carboniferous periods. The boundary between the Carboniferous and the Permian above it continues to be unsettled.

The lower boundary of the Carboniferous Period is defined by the appearance of certain mollusks, but the evidence is inconclusive and disputed. There is no distinguishing break in life-forms going from the Carboniferous to the Permian; thus, the upper limit of the Carboniferous Period has no generally accepted definition.

By contrast, the upper boundary of the Permian Period, although not precise in some global areas, is characterized by one of the greatest extinctions of plant and animal life in Earth's history, a worldwide destruction of species that brought to a close the Paleozoic Era. There is no consensus as to the cause of this great destruction of life.

Mesozoic Era. Determination of the upper boundary of the Triassic has been difficult, remaining inconclusive. The same is true for the Jurassic, which required an explicit (and subjective) agreement as to its placement.

Like the Permian, the Cretaceous is characterized by wholesale extinctions, whereupon the Mesozoic Era ends.

Tertiary Period of the Cenozoic Era. The Tertiary Period is subdivided into five epochs: Paleocene, Eocene, Oligocene, Miocene and Pliocene. These divisions were determined by the percentage of living species represented in their fossil content. The upper boundary, Pliocene/Pleistocene, is extremely nebulous, as the Pleistocene is not represented by layers of sedimentary rocks typical of earlier times.

"Unfortunately, few . . . boundaries seem near [a] stage of international acceptance and it will take many years before the entire stratigraphic scale is finally and irrevocably defined . . . "[4] writes one authority.

In the Standard Column, however, the ambiguous boundaries do not look any different from those which appear to have a basis in physical reality. Moreover, all are inscribed with dates that seem to be unquestioned (carved in rock, so to speak). But how certain are these dates?

The dates, like several of the boundaries, rely on guesswork. Most radiometric dates associated with the Column are obtained from igneous rocks that intruded into sedimentary layers while still molten. Ages of the two types of rock cannot be precisely correlated, nor is radiometric dating itself without its critics. Precise dating of Cambrian events is precluded by the lack of appropriate minerals in most Cambrian rocks.

The presumed rate of deposition of certain sequences of strata thus continues to be a major factor in determining the divisions of geological time. This assumption about the rate of sedimentary deposit and rock formation is based, of course, on modern observations under the umbrella of uniformitarianism; it disallows consideration of extraordinary circumstances. To admit the possibility of rapid deposition in a context of global catastrophe would endanger the foundations of many fields of science. But perhaps those foundations deserve more scrutiny than their defenders are willing to allow.

In the context of uniform process versus the extraordinary, comment is in order on the currently popular concept of *plate tectonics* as this relates to "continental drift."

Alfred Wegener proposed in 1912 that the world's continents were at one time joined together. He envisaged a supercontinent, *Pangea*, that began to separate during the Mesozoic Era. His hypothesis was based in large part on the apparent geographic fit of eastern South America and western Africa.* More recently, it has been suggested that there were originally two land masses. A competing idea holds that a single land mass, Pangea I, split during the Early Paleozoic into two. These two land masses, whatever their origin, reunited into Pangea II near the end of the Paleozoic, then underwent a final rifting that began in the Triassic Period of the Mesozoic Era.

After falling into disfavor and being ignored for a number of years, the idea of continental drift rose like a phoenix from its ashes when a new concept emerged, that of plate tectonics.

Instead of involving a "drifting action" of continents, continental drift relies on the idea of gigantic lithographic "plates" -- plates that rotate, translate and are constantly renewed by material pouring from the interior of the earth onto the ocean floor at "spreading zones." Slowly the plates spread from their source, and slowly they are drawn back into the interior, "subducted" at continental margins, to start a new cycle. As Charles R. Pellegrino and Jesse A. Stoff explain in *Darwin's Universe: Origins and Crises in the History of Life*:

> On the Mid-Atlantic Ridge, where part of the system that fractured Pangea is pushing Europe and North America apart at a velocity of several centimeters per year, huge convective currents in the Earth's mantle are bringing hot rocks and basaltic lava up to the surface. At the submarine ridge, and also in places like the Galapagos Rift, the Earth is releasing internal heat. The Mid-Atlantic Ridge is actively producing new sea floor and pushing it outward from the source.[5]

Pellegrino and Stoff do not continue through to the subduction of

---

* See further comments on Wegner under his name at the end of this chapter.

the plates, but suffice it to say that subduction is a geological process "in which one edge of crustal plate descends below another."[6]

As the Pellegrino-Stoff account suggests, the movements and cycles are conjectured to be empowered by plumes of heat rising through the mantle of the earth and gently urging the plates onward, the continents riding on top. The power source and, indeed, the plate movements themselves are at best hypothetical.

There is compelling evidence that the continents *were* united at some time in the past. In addition to geographical match-ups (of which there are several), former contact is suggested by certain distributions of glacial deposits and of plant and animal fossils. The remains of some creatures are found only in South America and Africa; these species apparently could not have gotten from one area to the other unless the habitats were connected as a single land mass. Certain rocks have been matched along the coasts of these and other continents.

There is other data that contradicts the notion of drifting land masses. Fossils of Paleozoic marine invertebrates -- which presumably lived on the continental shelves -- are interpreted as showing a pattern of climatic zoning by latitude similar to zoning seen today. The presence of these fossils, reflecting a seeming stability, would argue against any latitudinal migration of continents.

The contradiction, however, falls only into the camp of the uniformitarians; it does not impugn an argument for a separation of once-connected land masses. The positive evidence supports the concept of a land mass that split, but in no way does it attest to a separation by *slow drift* stretched out over millions of years. Drift is *inferred* from modern observations of continental movements, which supposedly range from one to 15 centimeters per year. As the *Time-Life* editors of *Continents in Collision* write, ". . . [G]eologists . . . believe that the same forces that widen the [African Rift] valley by about a millimeter per year have shaped most of the earth's features."[7] On the basis of this minuscule movement presently observed, and in the face of contradictory fossil evidence, contemporary geologists conclude that convective activity in Earth's mantle gradually tore Pangea in two.

But how can the gradualist look at a nearly static contemporary condition and dogmatically assert that the forces of geological change have never been any different? If I crush a stone with the sharp blow of a hammer and hand you the remains for observation, would you say that the stone from which they came could have been crushed only in a vise?

Such is the uniformitarian response to the geological record.

Since a different response is called for, perhaps a brief flight of fancy might be indulged.

Thinking the Unthinkable

Soon after the publication of his *Worlds in Collision*, Immanuel Velikovsky was accused of creating new forces in nature to suit his needs. That same year, J. H. Oort suggested (speculated) that there exists a diffuse cloud of gas, dust and comets -- gravitationally part of our solar system -- some 40,000 times as far from the sun as is the earth. Oort went on to suggest HOW MANY comets it contains, HOW it ORIGINATED, and HOW it RELEASES THE COMETS that we observe passing through our part of the solar system. Oort's hypothesis is the most popular idea today on the origin of comets *even though it does not rest on a shred of observational evidence.*

I once related to a professor of astronomy how Velikovsky had concluded from an extensive analysis of human records that Earth had been disturbed in its motion by a planet-sized comet some 3,500 years ago. The professor responded with a patronizing smile and challenged, "But a comet that large has never been observed." In turn, I smiled and asked, "When was the Oort Cloud last observed?" He smiled again but had nothing more to say.

Comets pass through the solar system on highly elliptical paths, making possible their passing near the orbits of planets. Should one pass near enough to a planetary body, the comet's orbit would be altered -- and altered in such a way that the probability of a second encounter might be greater than that of the first. If the two bodies had comparable masses, both would experience a change in their motions.

The initial chance of one of the masses being disturbed by the other would not depend simply on the probability of their reaching the same point in space at the same time. It would rest on the probability also of some degree of interaction of their respective gravitational and electrical fields.

Earth makes a complete rotation on its axis in approximately 24 hours. This motion translates into a linear velocity of about 1,000 miles per hour at the equator. The angular velocity (360 degrees in 24 hours) is constant everywhere on the earth, but the linear velocity decreases poleward. If other relatively minor movements were not present, linear velocity would be zero at the poles. It likewise would decrease beneath the surface, reaching zero at the axis. All motion on Earth's surface -- that of the atmosphere, the waters, the moving creatures -- is superimposed on this continuous rotational displacement.

Newton's First Law of Motion tells us that any moving object will attempt to move on the same path until disturbed or prohibited by an external force. This fact we all understand even if we have never heard it expressed. If you are riding in a speeding automobile that crashes to a sudden halt, you are likely to continue moving -- right through the windshield if not restrained. This tendency, of course, is less noticeable during normal braking.

The greater the mass and velocity, the greater the difficulty and trauma of stopping.

Suppose, now, that a giant body should come just close enough to the earth to retard, slightly, its rate of rotation. What would happen as a result? If a rotational deceleration of Earth were sudden enough, it would wreak havoc on Earth's surface -- a nightmare of horrifying proportions. Gale force winds immediately would begin to blow from west to east, continuing to blow for hours that stretched into days. Tornadic activity would quickly develop. Trees would be uprooted and structures destroyed.

Seas would recede from eastern coasts and encroach on western coasts. Sea levels would drop catastrophically in one region and rise in another as the waters began to move from west to east. Gradually the ocean waters would gather into a heap, moving ever eastward as they piled up, then would be swept across much of the world's land mass, crashing down far inland. Then they would recede westward, rolling back across the beds whence they came, inundating the land masses from the opposite side.

Upon approach of the waters to the land, all manner of sea creatures and oceanic detritus would be swept along and deposited across the continents, the load being sorted as moving waters always sort debris, the displaced waters tapering inland to filter finer and finer sediments until they thinned to nothing. Meantime, terrestrial animals and plants would be caught up in the cataclysmic flood, then unceremoniously dumped, along with uncountable tons of terrestrial detritus, on top of the oceanic layer. Remains would be scattered and mixed, ready to fossilize and to make uncertain the true habitats of all the life-forms when they were living. Alternate layers would accumulate as the waters rocked back and forth until finally they settled into their beds -- whether old beds or new.

Depending upon how quickly Earth's rotation slowed, equatorial sea levels might drop and those near the poles rise; or waters might rush first poleward, then back toward the equator, then poleward again until they settled, remaining lower in equatorial regions. Plant and animal remains would be scattered northward and southward. The same change in centrifugal force causing this poleward displacement of the oceans would create further storm and turmoil in the atmosphere.

The monolithic crust of the earth would crack -- later giving the impression of being composed of multiple "plates"-- with tremendous heat escaping from below the fissures. Lava would pour forth to incredible depths on land and in the seas, and myriad volcanoes would begin to spew forth rocks and gases from the depths of Earth.

As the layers of sediment continued to accumulate one on top of another, unrelenting horizontal forces would push the newly deposited layers of earth into heaps of mountain ranges, which would be folded and metamorphosed in the process. Older sedimentary rocks would be crumpled and thrust skyward by the same forces. Continents would be rifted along faults or lines of weakness and displaced many miles from where they once rested, a displacement made possible by the underlying molten or semimolten crust.

Given such planetary disarray, isolated types of plants and animals interred during the havoc would incur the possibility of being assigned by paleontologists and geologists to periods of time other than their own. "Evolutionary dating" would place them in an earlier time than some of their "more advanced" contemporaries. Thus, a fictitious time unit would be created to accommodate them.

Plant and animal remains from adjacent regions or land masses, carried by unimaginable hydrologic forces, would be buried in assemblages frequently unrepresentative of their living conditions, habitats or habits. Some would be removed from their dwelling places. There also would be areas where different types of life-forms overlapped in their burial -- areas that would separate regions of the earth where each type was buried alone.

It is easy to see that, for two similar but geographically separated life-forms thus distributed in death, their remains might subsequently be read as "an evolutionary sequence" in the fossil record.

There also could be some mixing, to a lesser degree, of forms that had been buried and fossilized in earlier upheavals, with further confusion of the record. Erosion and exhumation would be almost concurrent with deposition and reburial.

The natural electrical activity associated with atmospherics, volcanism and earthquakes of this magnitude would be phenomenal; its effects on the genetic structures of plants and animals, unpredictable.

Subterranean heat would raise the temperatures of the oceans in the equatorial and temperate regions. Evaporation would be rapid. In turn, this speedy evaporation would draw heat from the polar regions. The evaporated water would condense and precipitate at the poles in the form of snow and ice. Areas of polar ice would grow rapidly.

The consequences of an axial tilt of the earth could be much more severe than described above, since they would involve a change in the direction of movement of the atmosphere, the seas and the land masses relative to Earth as a whole. (This change might be moderated in places, however, depending on the direction of the tilt relative to the positions of particular geographic locales at the time of occurrence.) Prolonged interaction with a celestial interloper could result in continual, unpredictable irregular movements, each one a renewed assault on Earth's surface. An exchange of planetary materials would compound the destruction as rocks and debris rained from the sky.

Drastic changes could continue for days, weeks, months, even years. Reverberations might continue for millennia as the earth readjusted internally to a new geological equilibrium.

The Geological Record

The Geological Column presents the picture of a very orderly, gradual evolution of life on Earth. But the assumed evolutionary process dictates the interpretation of the geological record that is reflected in the Column. The circular logic runs thus: The record proves gradual evolution; gradual evolution is the basis for reconstructing the record.

As noted above, a perceived temporal, or evolutionary, separation of fossil types might arise from a misapprehension of data actually produced through a geographical separation confused by catastrophe.

The Paleozoic Era. Beginning with the Paleozoic Era and considering it as a whole, close examination shows that, with only two exceptions, the boundaries of its established subdivisions are not related to discrete physical conditions. But one of these exceptions is profoundly significant.

In North America there proves to be a disconformity in the Carboniferous Period that divides it into the Mississippian and the Pennsylvanian. This disconformity appears to extend to parts of the Old World, but it is not a universal phenomenon.

In contrast, the upper boundary of the Permian Period, which marks the end of the Paleozoic Era, is characterized by a worldwide extinction of many types of plant and animal life. Speculations on the cause of this extinction abound, but orthodox science has not been able to explain it in a way that is acceptable, even to all of those residing in the halls of orthodoxy. When sufficient evidence is considered, however, the cause should be obvious: wholesale catastrophe.

The Paleozoic rock sequence does not represent the Paleozoic Era in the sense usually urged. Rather, it represents the time during which the Paleozoic Era shook with its death throes -- years in length, probably, but not MILLIONS of years. At some unknown time in the past, a paroxysm of nature occurred, one that may have been initiated by an immense extraterrestrial force jolting our planet to the very core, decimating its biosphere, and restructuring its crust. The first world of Ancient Life -- a world of indeterminable age -- was to be no more.

The disruption of the Paleozoic World is to be envisioned as resulting from the gravitational and, perhaps, electrical interaction with an intruding planet-sized body. Such an encounter might have happened previously, but if there was then no life on Earth to become entrapped in the crust as fossils, effects of the earlier encounter very likely would be hard to recognize today. Precambrian sediments, which remain largely a mystery, could conceal a cataclysm. By contrast, the crustal imprint of the Paleozoic Era is recognized by scientists the world over.

What does that imprint reveal? Reprising the previous account -- in terms, now, of the Geological Column -- paints this picture: As Earth slowed on its axis or was rocked in its orbit, the oceans spilled out of their beds, carrying marine materials far inland. The first layers of deposits have come to be known as representing the Cambrian Period. As these layers were deposited, buckling of the crust began. The sediments were quickly folded in many places and soon were buried by the actions of the next cycles, the Ordovician and Silurian. As would be expected in such an upheaval, not all geological "periods" (rock systems) would be represented in all parts of the world.

As the oceanic intrusions continued inland and turbulence

increased, terrestrial creatures were caught up and entombed in subsequent sedimentation. These more recent beds are now perceived as later periods of "geological time." Indeed, the final result for neo-Darwinism is an imagined gradual evolutionary sequence of life-forms, from simple marine organisms in the beginning of the Paleozoic to complex terrestrial vertebrates prior to its close.

Much of the hypothesized geography of past eras is based on occurrences of remnant magnetism in igneous rocks. This magnetism is the source of information for studies of shifting magnetic poles and "continental drift." When igneous rocks cool and solidify, the magnetic domains within individual minerals align themselves in the direction of any magnetic field that might be present. This is usually assumed to be the magnetic field of Earth, and thus the rocks create (so the logic goes) a permanent record of how Earth's magnetic field was oriented when the rocks cooled. But if a stronger magnetic field, belonging to an external agent, happened to be present during cooling, the remnant magnetism would be worthless for ascertaining either the location of the magnetic poles or the positions of the continents.

The geography of the Paleozoic is thus highly uncertain if it is the result of cataclysm. Cardinal directions of the earth's surface at that time likewise would be speculative. What can be suggested is that as the cataclysm continued, marine inundations repeatedly assaulted the continental areas, from both east and west. There were north-and-south migrations of the shoreline of northern Asia. These motions would be consistent with what would be expected from changes in global rotation.

Reading the geological record in light of the foregoing cataclysmic assumptions, what else seems to have happened? The entire record of the rocks reads differently depending on the set of assumptions one brings to the reading task. This being true, a set of assumptions that is the complete antithesis of uniformitarian assumptions could be expected to illuminate a different Earth history than that of standard geological texts.

In fact, the reading offered here calls into question almost every conclusion of orthodox geological science.*

---

* Specific references to uniformitarian assertions presented in this chapter are not cited because these assertions can be found in any standard text on historical geology;

Geology has not only its own assumptions, but its own lexicon. "Transgression" and "regression" refer, respectively, to inundation of land masses by the oceans and to retreat of the waters following transgression. Most geologists believe that transgressions and regressions of the seas result, respectively, from the subsidence and uplift of a land mass, such as that attributed to North America during the late Paleozoic, or from a change in sea level. The thesis presented here is that changes in continental elevation or sea level were not the cause of the flooding by the seas; rather, a sudden change in global rotation flung the seas landward. The opposing encroachments (i.e., east-west) thus were not concurrent; they alternated. (1)

As the waters retreated, landlocked remnants formed large lakes -- scattered "inland seas" -- seas that for orthodox geology came about through glaciation.(2) Geologist Derek Ager has recognized the misconceptions prevalent in this respect, maintaining that "[m]ost sedimentation in the continental areas is lateral rather than vertical and is not necessarily connected with subsidence."[8] Indeed, it is of a tidal nature, from a disrupted rotation of our planet.

Mountains apparently began to rise as early as the Silurian Period (or wave of change). Chains appeared in eastern North America and in western Europe. Ultimately, two waves of mountain uplift encircled the globe.(3) Rock throughout the earth became progressively stressed. Temperatures rose. Lava poured from volcanoes and from gaping breaks in Earth's surface. The inland seas quickly evaporated as they were warmed by the overheated crust, leaving behind in their former beds large areas of evaporites (halite, gypsum).(2) Some of the mountains began to suffer erosion even as they were rising. Pounding waves poured over their tops, carrying the eroded material far inland, sorting and dropping their earthen cargo along the way. Huge wedge-shaped deposits resulted, stretching miles across continents.(4)

Before the end of the Permian, violent volcanism was extensive. Mountains continued to groan and grow. As the crust warped, whole continents shifted, taking on new configurations. Many forms of life disappeared forever. And a "geological age" came to its end.

---

however, a brief bibliography of works providing full documentation of my remarks is appended. Numbers in parentheses are keyed to specific events or phenomena identified in the geological record, as tabulated at the end of this chapter.

The origins of the Paleozoic ecology will always be a mystery. Events were not recorded (geologically) as they unfolded, as uniformitarians have led the world to believe that they were. Evolutionary development cannot be inferred from Paleozoic sediments because many of its parts (so-called systems) are virtually contemporaneous. Moreover, there is no geological record underlying the Paleozoic showing an earlier order from which it could have arisen. Whatever the origin, the Paleozoic apparently had a full spectrum of life as we know it, everything from microbe to reptile, long before its termination, except that fur and feather belonged to an age yet to come.

The Mesozoic Era. Subdivision boundaries of the Mesozoic are as elusive as those of the Paleozoic. And, as before, these subdivisions or periods are to be viewed, in a cataclysmic context, as being close together in time. Like the close of the Paleozoic, the Mesozoic upper boundary is marked by extinctions on a global scale, extinctions that uniformitarianism does not satisfactorily explain. Preserved in the rocks is the close of an apparent second phase of life and topography in Earth's history -- a phase known as the Age of Dinosaurs. Some of those dinosaurs were the largest land dwellers ever to live.

Our planet undoubtedly was a long time recovering from its Paleozoic battering, but like the phoenix, the mythical bird that rose to new life from its own ashes, it learned to thrive in a new form. And the biosphere took on an entirely new aura. Born somehow of cold-blooded creatures of the previous age, there arose animals of "hot blood," the beneficiaries of a biological mechanism called *homiothermy* (the internal maintenance of a specific body temperature). Some of these wore coats of fur. Others wore feathers. And these newly created mammals and birds assuredly were not so insignificant during that age as a scientific majority presently believe. Many of the dinosaurs were warm-blooded (first suggested by Velikovsky in 1941[9]), whether or not they had furry coats. Moreover, the often hellishly depicted airborne pterosaurs did have fur, as attested by their fossils,[10] and thus were not some kind of cold-blooded "flying lizard." Many sea creatures, such as the ichthyosaurs, probably were warm-blooded also, as are porpoises today, both of these creatures bearing live young. Another new arrival was flowering plants.

If a marauding planet-size body was indeed responsible for destroying the Paleozoic world, then -- barring the intruder's own destruction or some disturbance by a third body -- a second challenge of Earth by the marauder would not be out of the question: Such are the physics of orbital mechanics. Whether by the same hand or not, the geological record makes clear that global cataclysm brought another world age to its end -- just as before.*

When the second cataclysm struck, the wounds of the Paleozoic had healed, but now Earth was again wrenched on its axis. Once more the seas spilled out as the rotation was disturbed. The supercontinent began to rift. The Appalachians underwent faulting from Nova Scotia to the Carolinas. On the other side of the world in China, mountains buckled upward. In western North America, the Rockies and other chains began to appear. (5) All around the world the seas rose to cover much of the land, then withdrew; another ancient inland sea came and went within North America. (6)

The retreat of the waters brought the most extensive and prolonged flooding that the earth had experienced; continued stress forced further growth of mountain chains, and volcanoes coughed up their gases and poured out their lava. Lava engulfed huge areas and solidified to depths of thousands of feet. (7) The seas ran northward and southward as Earth's spin changed and attempted to stabilize. Before it was over, almost all of Europe was covered. A seaway connected the Gulf of Mexico and the Arctic Ocean. The Atlantic Coast fell victim to a "fluctuating shallow sea." South Africa, India, western and northern Australia, and South America all experienced three major transgressions of shallow seas.

Finally, the seas found their proper basins and a calm returned. The

---

* Chapter I refers to the thesis of Luis and Walter Alvarez that Earth probably was struck at the end of the Cretaceous Period by a Manhattan Island-sized asteroid and that this collision may have had a great deal to do with the extinction of dinosaurs. Pellegrino and Stoff note that the "anomalous iridium levels [found by the Alvarezes in rocks] . . . point almost like a finger toward the North Sea, and there, nearly 325 kilometers (200 miles) east of Dundee, Scotland, lies another anomaly: a 250-meter (820 foot)-deep depression on the sea floor known as the Devil's Hole. The 'hole' forms an almost perfect circle. measuring over 80 kilometers (50 miles) in diameter."[11] This hole may be the crater from the impact of the asteroid. The Alvarezes argued that "ten years of darkness" and severe changes in climatic conditions were caused by the asteroid collision, leading to the mass extinctions at the end of the Cretaceous period. The Alvarez hypothesis gives no credit to Velikovsky's priority for "collision" explanations of geological facts; it also does not come close to propounding the degree of violence that actually engulfed Earth at different times in the past.

Mesozoic Era ended in a reign of death. The dinosaurs, as well as myriads of other creatures, were entombed in the crust, curiosities for a future time.

Given the scarcity of natural boundaries, one fact is highly interesting: The Paleozoic is divided into twice as many periods as the Mesozoic. Is this difference related to physical phenomena, or is it an artifact of evolutionary theory? Perhaps a more protracted upheaval would result in a broader range of gradation. Equally probable is another possibility: The Mesozoic is viewed by uniformitarianism as having a full range of life-forms from its very beginning, from "simple" to "complex." By contrast, the Paleozoic is viewed as "starting from scratch." I cannot escape a suspicion that the Darwinists simply needed more time periods to accommodate "the progression of lower forms of life." Thus, these ladder-like Paleozoic divisions are a part of the official Column.

The Cenozoic Era. The Cenozoic Era is divided into two periods, the Tertiary and the Quaternary. The Quaternary is further divided into the Pleistocene and Holocene Epochs, the former popularized as the time of the Ice Ages. The Holocene is the epoch in which we live today.

After the devastation ending the Mesozoic, the world again was a long time recovering. But a new world dawned as before, a new world and another new order of life. Most of the many strange citizens of the Mesozoic had disappeared, but in their passing they gave rise to descendants unlike themselves -- they became vessels of a process that engendered novel progeny from the very forces of destruction. The new age was one still populated by giants, but giants on a smaller scale than those that came before. These new colossuses were mostly furry giants, some feathered. Albeit bizarre, they would look vaguely familiar today. Included among these were the first primates. Among mammals in the ocean were numbered the whales and porpoises, transformed descendants of bygone inhabitants of Mesozoic seas. The notion of their having gradually evolved from land mammals is not supported by the evidence; their progenitors were ocean dwellers of similar forms.

The variety of Tertiary mammals was phenomenal. Indeed, that period has come to be known as the Age of Mammals. But it, too, was destined to end catastrophically: A third time celestial forces gripped our planet. Again the seas fled their boundaries, covering much of Europe and Africa and then retreating, bringing eight major transgressions of North America from the Gulf. More of the crust was buckled into mountains: Europe, Asia, New Zealand. A final time the continents slipped, attaining their present positions. Lava poured through clefts in Earth's surface, some of the flows again reaching (and solidifying to) thousands of feet in depth. Countless species of animals were doomed to extinction.

The variety of new life that was born in the midst of the Tertiary disaster was not as extensive as had been the case in earlier times. Yet from our perspective as a species, this new life was certainly significant. From unknown progenitors there came *Homo erectus*, forerunner of *Homo sapiens*. Next most interesting, perhaps, was the still-mysterious *Australopithecus*, also of erect stature, but very far from being human. This creature has no equivalent today.

Our overview of this long cycle (?) of catastrophe must close with a few observations on the beginning of the Pleistocene "Ice Ages," the next epoch to follow -- first epoch of the Quaternary Period.

As the turmoil of the Tertiary dragged on, Earth was again overheated, as evidenced by the vast amounts of volcanism. The waters in the equatorial and temperate regions likewise rose in temperature and evaporated rapidly. As a result, heat was withdrawn from the polar regions. Atmospheric vapors precipitated at the poles in the form of ice, and the polar caps expanded, bringing with them the onset of the first Pleistocene ice age. But as the earth was released from its torment, it once again stabilized. The new climate would not sustain the extensive ice cover that had formed, and it began to retreat, leaving in its wake the First Interglacial Period.

This cataclysmic view of geological time would not be complete without a sidelong glance at standard hypotheses about this first and all subsequent Pleistocene ice ages. While the retreat of the first Pleistocene ice cover appears to have been slow, the rapid initial formation of it is enough to defeat all of the calculations upon which the accepted time scale is based.

Moreover, subsequent advances and retreats of the ice also were associated with global disturbances -- dictating a reassessment of Quaternary chronology.

Conclusion

The Standard Geological Column seems to imply that Earth is undergoing a never-ending burial. Uniformitarians would stress that the burial is not universal at any single point in time and that continual eruptions of new matter (creating new geological features) occur independently as well as concurrently. Nevertheless, observations of such "interments" as do occur -- volcanoes, earthquakes, regional flooding, etc. -- along with the recovery from many of their effects, should make abundantly clear the fact that the geological record cannot provide a continuous picture of Earth's history. Global information simply is not frozen into stone in the course of everyday existence. Noncatastrophic preservation of fossils is *relatively* insignificant. Even with uncounted millions of years to accumulate a record through everyday processes, the quantities of accumulations in the fossil record would not compare even remotely to those that actually exist.

Theorists of a catastrophic geological past (like myself) recognize that uniformity has been the *geological operative process* for most of Earth's history. That is, through the actions of weathering, erosion, transportation and deposition, the face of the globe is being altered constantly and ever so slowly. But forces of a magnitude actually to sculpture the major features of our planet are seldom seen -- *never* in our lifetimes. Only because of these violent intrusions can anyone talk about a "geological column"; without them, there would be no column. Therefore, the schematic column should be modified to show a cyclical arrangement of uniformity-catastrophe (see the revised "Column" in the "Afterword").

Historical geologists have brought to light many key events in Earth's long history, some never even suspected during most of humankind's existence. However, their reconstruction of the context, time scale, and causes of events has suffered from the constraints of uniformitarianism.

The present is *not* the key to the past, and the assumption that it is has kept much of the past hidden from us. Not only Earth itself, but, as shall hereafter appear, the many forms of terrestrial life that dwell upon it have progressed through time in a way hardly imagined by orthodox science.

ALFRED WEGENER

"Many scientists tried to explain why the [initial] reaction to Wegener's theory was so predominantly hostile. Some argued that he was simply premature, too far ahead of his time to be accepted. Some suggested that any scientific hypothesis that contradicts the prevailing view will be set aside at least for a time in the hope that it will turn out to be invalid. But Wegener was not just set aside: His ideas were ridiculed and relegated to the realm of fantasy. Many of his colleagues preferred to close their minds to obvious contradictions in the prevailing theories about the origin of the earth, rather than consider Wegener's proposals. . .

"In the final analysis, Wegener may have been rejected because . . . he was an outsider. He was neither a geologist, nor a paleontologist, nor a biologist, yet his hypothesis trampled across all those fields, crushing underfoot long-held and cherished notions. 'The trouble must partly have been,' Oxford geologist Anthony Hallam would write four decades later, 'that he was not an accredited member of the professional geologists' club.' We of course now see it as a positive advantage that Wegener had not been brainwashed by the conventional wisdom as a student. His position was an advantage because he had no stake in preserving the conventional viewpoint. He was not an amateur, but an interdisciplinary investigator of talent and vision who surely qualifies for a niche in the pantheon of great scientists."[12]

One day this quote undoubtedly will be applied to Immanuel Velikovsky.

THESE ITEMS KEYED TO TEXT

1. Paleozoic Transgressive/Regressive Cycles in North America

Sequence Name	*Time*
Sauk	Late Precambrian-Ordovician
Tippecanoe	Ordovician-Early Devonian
Kaskaskia	Devonian-Mississippian
Absaroka	Pennsylvanian-Permian

2. Paleozoic Examples of "Inland Seas/Evaporite Deposits"

Name	*Location*	*Time*
Absaroka Sea/Paradox	Western U.S.	Early Permian
Zechstein Sea	Northern Europe	Late Permian

3. Paleozoic Orogenies (Mountain Building)

Cycle			
	Name of Orogeny	*Mountains or Location*	*Time*
Caledonian			
	Taconic	Northern	Ordovician-Silurian
	Caledonian	Western Europe	Silurian
Hercynian			
	Acadian	Northern	Devonian
	Antler	Nevada to Alberta	Late Devonian
		Central Europe	Carboniferous-Perm.
	Allegheny	Southern	Mississippian-Perm.
	Somona	Western Nevada	Mid-Permian

4. Paleozoic Example of "Clastic Wedge"

Name	*Location*	*Time*
Queenston Delta-500 mi across	N.E. North America	Late Ordovician

5. Mesozoic Orogenies

Cycle			
	Name of Orogeny	*Mts or Location*	*Time*
Laramide			
	Nevadan	Western No.	Late-Jurassic-Mid-Cret.
	Sevier	Rockies	Mid-Jurassic-Cretaceous
	Laramide	Western No.	Mid-Jurassic-Cretaceous

6. Mesozoic Example of "Inland Sea"

Name	*Location*	*Time*
Sundance Sea	Western North America	Jurassic

7. Mesozoic Example of Profuse Lava Flow

Name	*Location*	*Size*
Deccan Traps	India	193,000 sq mi/10,500 ft thick

REFERENCES

1. Immanuel Velikovsky, "On Saturn and the Flood," *KRONOS* V-1 (Fall 1979), p. 7.

2. Velikovsky, *Worlds in Collision*, Doubleday (Garden City 1950), passim.

3. Ibid.

4. David G. Smith, "Historical Geology: Layers of Earth History," *The Cambridge Encyclopedia of Earth Sciences*, ed. by Dr. David G. Smith, Crown Publishers (New York 1981), p. 390.

5. Charles R. Pellegrino and Jesse A. Stoff, *Darwin's Universe: Origins and Crises in the History of Life*, Van Nostrand Reinhold (New York 1983), p. 174.

6. *The American Heritage Dictionary*, 2nd College Ed., Houghton Mifflin (Boston 1982), p. 1,211.

7. Russell Miller, *Continents in Collision*, Time-Life (New York 1983), p. 31.

8. Derek V. Ager, *The Nature of the Stratigraphical Record*, 2nd ed., Halsted Press (London 1981), p. 66.

9. Velikovsky, "Were All Dinosaurs Reptiles?" *KRONOS* II-2 (November 1976).

10. Edwin H. Colbert, *Evolution of the Vertebrates*, John Wiley & Sons (New York 1980), p. 181.

11. Pellegrino and Stoff, op. cit., p. 153.

12. Russell Miller, op. cit., p. 57.

BIBLIOGRAPHY

Cambridge Encyclopedia of Earth Sciences, ed. by Dr. David G. Smith, Crown Publishers, Inc., Cambridge University Press (New York 1981).

Adrian J. Desmond, *The Hot-Blooded Dinosaurs*, The Dial Press (New York 1976).

Implications of Continental Drift to the Earth Sciences, ed. by D. H. Tarling and S. K. Runcorn, Academic Press (London 1973).

William J. Kaufmann, III, *Universe*, W. H. Freeman and Co. (New York 1985).

Bjorn Kurten, *The Age of Mammals*, Columbia University Press (New York 1972).

Harold L. Levin, *The Earth Through Time*, Saunders College Publishing (Philadelphia 1983).

Donald H. Menzel, *Astronomy*, Random House (New York).

George Gaylord Simpson, *The Geography of Evolution*, Chilton Books (New York 1965).

Immanuel Velikovsky, *Earth in Upheaval*, Doubleday (Garden City 1955).

CHAPTER VI

The Path of Life: A History of Interruption

In his book *Creation-Evolution: The Controversy*, R.L. Wysong, like many of his creationist predecessors, has made an excellent case against Darwinism. While the proposals that he offers in support of creationism rest on the misinterpretations enumerated in previous chapters of this book, Wysong, unlike many of his fellow apologists, leaves room for a reconsideration of the ground on which his own ideas rest. "When and if another alternative is defined," he writes, "we'll consider it."[1]

We shall see.

One chapter of this book challenges the evolutionist concept of speciation by natural selection. Contrary to all the dogmatic hoopla, this supposed process is fictitious.

Another chapter challenges the creationist's interpretation of the Book of Genesis. Misinterpretations of biblical texts have undermined much of the creationist position; partly as a result, the creationist has fictions, too (e.g., the vapor canopy).

Only the rash will rush onto ground that angels fear to tread and try to say, once and for all, *here is the truth about everything.* There are of course certain known facts. The existence of the physical universe and the laws that govern it are facts. Nevertheless, we remain in what the

Romans called *mare tenebrarum* -- the sea of darkness -- with regard to the world. "The only solid piece of scientific truth about which I feel totally confident is that we are profoundly ignorant about nature," said marine biologist Lewis Thomas.[2] Seconding Thomas' humility can be liberating: It dispenses with the need to be or appear infallible.

That said, here is an examination of the bases on which ideas about speciation rest, together with suggestions for a new view of speciation.

The life-forms of Earth have changed through time, a fact attested by the fossil record. Since this record is the only kind of objective evidence available to us, some consideration is due to the basic fabric of the record -- to the context of fossil formation -- before probing into the nature of speciation.

The Origin of Fossils

If somehow we could spend a number of years beneath the oceans upon the floor of the sea, we might be able to see the initial part of the fossilizing process. As countless tiny life-forms die and are buried in the mud generation after generation, they accumulate in layers. After a long time, they might indeed be preserved as though stone, or even become stone. It is obvious from studies of some sedimentary rocks that many small creatures have died peacefully and dropped to their graves undisturbed. Similar processes might be observed in any marine environment. These are processes that are slow and never-ending. And the uniformitarian exclaims, "This is the way it all works!"

Now we leave our watery environment to observe elsewhere. What is happening to the land dwellers? We come across a dead antelope somewhere in Africa. Perhaps we can wait to see if it is going to start fossilizing. But no, a scavenger comes along and devours him and scatters his bones. We move elsewhere and see another animal die, followed by a fortuitous burial in a mudslide. Perhaps it will fossilize. But it is only a solitary carcass. If not for the scavengers and the isolated mudslide, both animals would have decayed long before fossilization could have occurred.

We continue to look, but in vain: We cannot find any significant occurrence of land dwellers, plant or animal, being fossilized. We naturally conclude that such things just do not happen.

But maybe, once upon a time, they *could* have happened. So we scout around the world again, but this time we dig. We are amazed at what we find.

Everywhere are the fossilized remains of plants and animals -- even those of giants. Predator and prey alike are jumbled together. Not only individual bones, but entire skeletons are sealed in the rocks. There are even mixtures of trees that grew at different latitudes. Digging, we find countless varieties of fossils by the countless thousands.

In Scotland, there is an area of perhaps 10,000 square miles brimming with fossil fish -- not the peaceful little sinking shells mentioned previously, but whole fish with fossilized soft parts. Their fins are outspread and their bodies contorted from the fear that surrounded their death.[3]

In the Arctic regions are found mammoths frozen in the ice with traces of delicate plant life in their teeth and stomachs.[4]

In 1976 the fossilized skeleton of a baleen whale was found in a diatomaceous earth quarry in California. Estimated to be some 80 feet in length (before excavation), it is one of the largest fossils ever found. But more astounding than the size is this: The whale was standing on end![5] As to the matrix in which it was encased, conventional wisdom informs the world: "Deposits of diatomaceous earth were formed by the slow [read 'millions of years'] accumulation on ocean floors of the shells of minute algae called diatoms."[6] The incongruity of this description and the whale's position has led someone to remark that this must have been the greatest balancing act in history. The alternative to this out-Barnum-and-Baileying of the circus acrobats is a vastly different process than the Darwinesque geologist expounds.

Apparently oblivious to the facts of his discipline, and without ever batting an eye, paleontologist David Norman tells us, "Rapid burial is most likely to occur in the sea, where a constant 'rain' of sediment falls on the sea floor. . . . Land dwellers such as the dinosaurs also get preserved by burial, but the carcasses have to be washed into a lake or into the sea."[7]

Another scientist raises doubts about this process. Fossilized tracks of a Jurassic horseshoe crab have been found with "the tracks terminating at the actual horseshoe crab that made the tracks! Is this an example of an animal walking along and suddenly, almost instantly, being entombed in sediment? [What else, pray tell?] Such graphic fossil occurrences tell us much about the animal, and also *provide us with cause to reflect upon the mechanics of fossilization itself.*"[8]

Clearly, all of these creatures, whether on land or in the sea, were destroyed and buried quickly, and by such violence that some were carried many miles and scrambled with others with which they possibly never had contact in life. The previous chapter gave a glimpse of the magnitude of this violence.

Occasionally, a myopic recognition is accorded by orthodox geologists to the apparent hand of catastrophe in some depositions. Writes Derek Ager:

> . . . we do from time to time find evidence, in all parts of the stratigraphic column, of very rapid and very spasmodic deposition in the most harmless sediments. In the late Carboniferous Coal Measures of Lancashire, a fossil tree has been found, 38 feet high and still standing in its living position. Sedimentation must therefore have been fast enough to bury the tree and solidify before the tree had time to rot. Similarly, at Gilboa, in New York State, within the deposits of the Devonian Catskill delta, a flash flood (itself an example of a modern catastrophic event) uncovered a whole forest of *in situ* Devonian trees up to 40 feet high. . . . Undoubtedly, comparatively sudden and very widespread events, such as major marine transgressions, did occur at various times during the earth's history.[9]

Through his description of a flash flood as "an example of a modern catastrophic event," Ager betrays his parochial concept of catastrophism. Trapped in uniformitarian assumptions, he falls short by several orders of magnitude of perceiving the scale of catastrophe necessary to write the fossil record found in Earth's crust.

In a rare instance of published humor, Velikovsky describes in his *Earth in Upheaval* a situation in Europe, contrasting the obvious with Charles Lyell's ludicrous explanation of "historic non-events." Velikovsky's persiflage-mingled critique is presented here in its entirety.

> **The Hippopotamus**
>
> The hippopotamus inhabits the larger rivers and marshes of Africa; it is not found in Europe or America save in zoological gardens where specimens of it wallow most of the time in pools, submerging their huge bodies in muddy

water. Next to the elephant it is the largest of the land animals. Bones of hippopotami are found in the soil of Europe as far north as Yorkshire in England. Lyell gave us the following explanation for the presence of the hippopotamus in Europe:

"The . . . geologist may freely speculate on the time when herds of hippopotami issued from North African rivers, such as the Nile, and swam northward in summer along the coasts of the Mediterranean, or even occasionally visited islands near the shore. Here and there they may have landed to graze or browse, tarrying awhile, and afterwards continuing their course northward. Others may have swum in a few summer days from rivers in the south of Spain or France to the Somme, Thames, or Severn [river in Wales and England], making timely retreat to the south before the snow and ice set in."

An Argonaut expedition of hippopotami from the rivers of Africa to the isles of Albion sounds like an idyll.

In the Victorian cave near Settle, in West Yorkshire, 1,450 feet above sea level, under twelve feet of clay deposit containing some well-scratched boulders, were found numerous remains of the mammoth, rhinoceros, hippopotamus, bison, hyena, and other animals.

In northern Wales in the Vale of Clwyd, in numerous caves remains of the hippopotamus lay together with those of the mammoth, the rhinoceros, and the cave lion. In the cave of Cae Gwyn in the Vale of Clwyd, "during the excavations it became clear that the bones had been greatly disturbed by water action." The floor of the cavern was "covered afterwards by clays and sand containing foreign pebbles. This seemed to prove that the caverns, now 400 feet [above sea level] must have been submerged subsequently to their occupation by the animals and by man. . . . The contents of the cavern must have been dispersed by marine action during the great submergence in mid-glacial times, and afterwards covered by marine sands. . . ," writes H.B. Woodward.

Hippopotami not only traveled during the summer nights to England and Wales, but also climbed hills to die peacefully among other animals in the caves, and the ice, approaching softly, tenderly spread little pebbles over the travelers resting in peace, and the land with its hills and caverns in a slow lullaby movement sank below the level of the sea, and gentle streams caressed the dead bodies and covered them with rosy sand.

Three assumptions were made by the exponents of uniformity: Sometime not long ago the climate of the British Isles was so warm that hippopotami used to visit there in summer; the British Isles subsided so much that caves in the hills became submerged; the land rose again to its present height -- and all this without any action of a violent nature.

Or was it, perchance, a mountain-high wave that crossed the land and poured into the caves and filled them with marine sand and gravel? Or did the ground submerge and then emerge again in some paroxysm of nature in which the climate also changed? Did the animals run away at the sign of the approaching catastrophe, and did the trespassing sea follow and suffocate them in caves that were their last refuge and became the place of their burial? Or did the sea sweep them from Africa, throw them in heaps on the British Isles and in other places, and cover them with earth and marine debris? The entrances to some caves were too narrow and the caves themselves too "shrunk" (contracted) to have been places of refuge for such huge animals as hippopotami and rhinoceroses.

Whichever of these answers or surmises is correct, and whether the hippopotami lived in England or were thrown there by the ocean, whether they sought refuge in caves or the caves are but their graves, their bones on the British Isles, as also on the bottom of the seas surrounding these islands, are signs of some great natural change.[10]

Missing Links

The standard geological column implying a gradual, orderly progression of biological change is a myth. Even in the most ideal situations, there is not to be found a uniform grading of any species into another. The gaps in the fossil record are real; the desired transitional life-forms are not.

The uniformitarians have had their P.T. Barnum to fabricate a Feejee Mermaid; now they need a Merlin to summon more convincing missing links from the unaccommodating rocks.

In the mind of the general public, the Lyellian and the Darwinian succeeded well over a century ago in inverting the Great Chain of Being. But the loops in the Darwinian's chain are not linked together; no loop is connected to the next loop "up."

According to G. G. Simpson, "This regular absence of transitional forms . . . is an almost universal phenomenon, as has long been noted by paleontologists."[11]

The introduction to the 1956 publication of Darwin's *Origin of Species* states: "What the available data indicated [in Darwins' day] was a remarkable absence of the many intermediate forms required by the theory . . . [and] the position is not notably different today."[12]

The "gap evidence" persists -- and it is real, observable evidence. Unfortunately, it has been observed but ignored, rationalized, explained away -- anything but faced squarely. What continues to seem remarkable is that the most damaging blow -- the almost-seminal error -- was committed by so dispassionate an observer of the geological record as Charles Darwin appears to have been. But Darwin came to his task clanking the chains of a philosophical spectre. He reasoned from an intellectual context that grew from Neo-Platonic assumptions about an unbroken chain of life, and he came up with a theory that demanded -- what else? -- an unbroken chain of life.[13]

In light of all the known and publicly recognized gaps in the record, the following surely must be one of the most absurd claims to be found in the pages of *Encyclopaedia Britannica*:

> Several writers have suggested that, in addition to the "microevolution" that produces species, there is also a process of "macroevolution" that, through a sudden and major genetic change, may produce a whole new organism, assignable to a different major taxonomic category from that of its parents and founding a new line; but no evidence from nature has been offered to support such a theory [i.e., the gaps are denied], which appears to run counter to the *wealth of information on natural selection*."[14]

Shades of the peppered moth! The *Encyclopaedia Britannica* passage reveals the calcification of the academic mind around a notion that grew, accidentally, out of the ancient assertion that "nature leaves no gaps." This idea was a major part of the Great Chain of Being, and when the Darwinian turned the Great Chain of Being upside down, he unthinkingly kept the Neo-Platonic notion of a chain without gaps. In fact, he was blinded, by the residue of a failed philosophical system, to the fact that his "chain" lay in disarray -- unconnected, unconnectable, thousands upon thousands of unbroken loops, none of which were "linked" to other loops. Poor Charles would have been better off with his Grandfather Erasmus' expansive nisus; that, at least, would have provided an "occult property" (a teleological force) to "explain" the gaps in nature.

But even a spiritual "striving" of nature for ever higher expression could not by itself explain the disarray of the fossil record. And aside from gaps and disarray, other anomalies of the record are frequently ignored by scientists and not revealed to laymen.

Nowadays, of course, ignoring such inconveniences is easy: Darwinian after Darwinian after Darwinian has laid down his deposits of sediment on the uniformitarian plains until the weight and pressure of all that detritus has made solid rock of the intellectual orientation of the geological establishment. Viking and Apollo space missions bring back startling new facts about our universe, and the uniformitarian but bends these facts tortuously to his preconceptions. His orientation being inelastic, he can look in only one direction. What hammer shall be large enough and swung with sufficient force to shatter *that* solid rock?

Between the Gaps

As already noted, there is other solid rock to be broken if a truly earnest re-examination of the geological record is to commence. That rock is the intellectual orientation of the creationist.

Despite the gaps between loops in the Darwinian "chain," the fossil record does not demonstrate that biological change has been as limited as most creationists suggest. According to them, every modern form has come from a line of fertility extending back to an original created type. This is a forced conclusion based on a specific *erroneous* interpretation of the first chapter of Genesis, as previously demonstrated. A rare confession of the uneasiness of the creationist position on "kinds" is found in J. W. Klotz: "There is no doubt that the fossil evidence does pose a real problem to those who would insist on a limited amount of change since Creation."[15]

A point offered in proof of creationism is the argument that the fossil record attests to the sudden appearance of fully developed life-forms. From *Creation-Research Society Quarterly*:

> . . [T]he fossil record, although claimed by some to be evidence for evolution, is in fact more indicative of special Creation. A well known book on geology, published in 1958, corroborated in a remarkable chart the Creationist position, although the book was not written from a Creationist viewpoint. It showed that many distinct kinds of creature, both plant and animal, appeared suddenly and separately in the fossil record, and remained separate, without any evidence of evolution into something else, until the present or until they became extinct.[16]

There is something to be said for this claim, but it should be evaluated in light of the creationists' other assertions.

Fossils are formed at times of physical upheavals; according to Velikovsky and others (including myself), the great bulk of all fossils are formed in this way; thus, the fossil record is a by-product of destructive processes that occurred well after the proliferation of the life-forms mirrored in the rocks. What should be clear is that all life-forms would appear in the record suddenly at the time of catastrophe, regardless of how they were created or developed -- even if they had experienced 100 million years of evolution.

If the creationists are claiming *successive* sudden appearances of new forms, they are implying the occurrence of some catastrophe between stages -- which counters their arguments for both a six-day Creation and a single cataclysm (the Deluge).

Since extreme variations are possible within some living species and since even males and females of other species bear little resemblance to one another, caution should be exercised in the interpretation of the fossil record. Size, morphology and other factors notwithstanding, neither creationists nor evolutionists can say whether an apparently newer life-form represents a degeneration, a complexification, or just a "neutral shift" relative to its presumed predecessor. Surely the "simple" came first, but our preconceptions cause us to stumble.

Biological transmutation is real. It is obvious through comparison of living forms to fossils. The cause and time frame (and sometimes direction) of that change are theory; they have not been demonstrated or verified.

Gaps Among the Living

In the 1958 edition of Darwin's *Origin of Species*, Julian Huxley wrote in the introduction: "Today, a century after the publication of the *Origin*, Darwin's great discovery of the *universal principle* of natural selection is firmly and finally established as the *sole* agency of major evolutionary change."[17]

What is there to make this "sole agency" the mainstay of evolution? What grand mechanism exists therein to generate new species? Exactly what does this "universal principle," natural selection or "survival of the fittest," explain?

Philosopher R. H. Brady has argued that despite various twists of language, natural selection means simply that the fittest species have survived. But the theory does not specify how fitness is to be determined. The usual answer to this embarrassing question is that fitness means leaving more offspring, but this is only a roundabout way of saying, "By surviving." It produces a circle: The surviving species survive because they are the fittest, and they are adjudged to be the fittest because they survive. There is no independent criterion of fitness, so the term means only that those who survive have survived. Obviously correct, such a statement is uninformative; it explains nothing and is useless.

In summary, the supposed process of evolution is "genetic mutation" and "survival of the fittest," repeated *ad nauseam*. Aside from the problems already cited, the scheme in many instances defies logic.

Consider the kangaroo, a marsupial. It is born prematurely, exceedingly small, unable to cope with the world it enters. It must immediately make its way to the security of its mother's pouch; if the pouch were not there, it would perish. Why is there a pouch? If the pouch had not been necessary -- "forced to develop" -- it would not have developed. But if it was necessary, what were kangaroo young doing in the meantime, prior to pouch development?

Survival of the fittest: In this case, perhaps a race of pouchless kangaroos competing with those having pouches. How did the pouchless race protect its young while competing? If the pouchless were to survive as a race, their young had to be prepared to meet a world without pouches -- but in that case, pouches would have been unnecessary.

Gould and Lewontin, as quoted before, have noted the fecundity of "explanations" by Neo-Darwinians: "[T]he rejection of one adaptive story usually leads to its replacement by another, rather than to a suspicion that a different kind of explanation might be required. Since the range of adaptive stories is as wide as our minds are fertile, new stories can always be postulated. . . ."[18]

In the case of the kangaroo, the Darwinian ministry might posit a two-stage process: (1) A race of kangaroos with pouches gained an advantage over pouchless kangaroos because ecological conditions favored the female adult that could dispense with a period of preoccupation with its young; (2) deteriorating ecological conditions then favored the pouched kangaroo that also could have a shorter gestation period, and the utterly defenseless premature young of today represent a retrograde state for the baby kangaroo but at one time, at least, represented a decided advantage for the adult female.

See how easy it is? Anything goes. Anything *might* have been the case. This forced that, and that forced the other. Or congeries of advantages for a new adaptational development led to its dominance. (The layman must always remember the extreme unlikelihood of any mutation, as observed in numerous experiments, proving to be favorable in any way at all.)

What is utterly apparent is that if kangaroo young always were as premature and ill-equipped for survival as they are now, the "system" -- mother, young, pouch -- had to be fully functional at the outset, else there would be no kangaroo. This is but a simple example; nature has myriads of others. An agile enough imagination can always supply adaptive stories -- an endless supply as each new story is thwarted by observed or newly discovered facts -- but the plots are strained and artificial.

The truth is that isolated, randomly mutating genes operating under gradualist skies are impotent to account for the complexity, organization, and interrelationships of life around us.

What was the Darwinist path that led to the metamorphic process experienced by the caterpillar? There are no stepping stones in achieving such a life cycle. Anything going through a "soup stage" had to be ready all at once.

Clearly a new paradigm is required for theories of origins.

Modes of Speciation

There appear to be two modes of speciation, one of which has actually been observed, neither of which involves gradualism. They might be referred to as restricted speciation and mass speciation.

Speciation I. Restricted speciation, involving a single pair of organisms and their offspring, is the phenomenon of *polyploidy*. This, in fact, is fairly common for certain classes of sexual plants and reportedly has been seen to occur in some lizards and frogs. It involves a doubling of chromosomes in the offspring, making the offspring unable to backcross with either parent; however, the offspring can breed with other organisms like itself. An author in *Encyclopaedia Britannica* gives an example from the plant kingdom:

> Genetic isolation is produced when an organism multiplies the number of its chromosomes (polyploidy). This is potentially important in producing hybrids that cannot breed with either parental stock because of the incompatibility of their sets of chromosomes with each parental species. A hybrid primrose formed by crossing *Primula floribunda* with *P. verticiliata* underwent a doubling of the number of chromosomes possessed by the parental species, which meant that each chromosome then had a compatible partner for the mechanism of cell division. This polyploid hybrid was then able to set fertile seed and develop into a plant with its own constant characters, whose offspring bred true. The new plant was intersterile with both its parent species and is *Primula kewensis*. . . .[19]

Speciation II. The Synthetic Theory postulates evolution through the almost infinitely slow transformation of one species into another (anagenesis) and through a branching from one species into two or more new species (cladogenesis). It disallows, however, the concurrent emergence of many species comprising a basic structural design, as grouped into higher taxa.

The species is central to the issue of this kind of biological change because it is the only natural taxon that might be conclusively identified. The higher taxa (order, class, etc.) are based primarily on presumed evolutionary relationships. Species belonging to the same genus are believed to be closely related (recent common evolutionary ancestor) even though they are not interfertile. Evidence for this is thought to be seen in animal behavior (recognition) and specific parasite preferences (e.g., all the species of a given genus).[20] Beyond the species, for which interfertility is the basis for identification (of living, sexually-reproducing species), this schema obviously becomes increasingly subjective for higher and higher taxa. I would suggest, however, that the members (separate species) of a true *natural* genus did arise -- suddenly -- from a single type.

When stripped of artificial uniformitarian constraints, the fossil record bears witness to just such grand-scale biogenesis. And it shows that these changes -- speciations -- came at specific times in Earth's history, during periods of great upheaval, great extinctions, and widespread environmental changes.[21]

Speciation originates in the sex cells of the parent generation. New chromosomal and genetic structures dictate(d) revisions in anatomical organization, development, physiology, morphology, neurology. We can only wonder to what extent parent and offspring differed, especially in those cases in which the newborn required parental care. Too much difference might cause parental rejection.

Rejection or not, a biological barrier was raised between an old line and a new one -- and raised in a flash -- producing a generation of new creatures not interfertile with their parent generation.

Given a sudden environmental change, it is possible that within the new environment, over a prolonged period, certain species, whether old or new, might be "fine-tuned" with respect to that environment (adaptation to the environment within species limits). Others might not survive. These changes in ecology, however, would not entail some form of "Darwinian survival of the fittest." The adaptive failures would be better described as the "rapid demise of the incompatible" -- most likely the first generation after the change. (This is a legitimate theoretical distinction. It is *not* just the other side of a Darwinian coin.)

Random In, Non-Random Out

Speciation certainly is not to be compared to spontaneous generation; it is not a bootstrap operation. But what could do it? Whatever the cause, it must be (have been) external to the affected organism. And surely the stimulus was a random product of nature.

Velikovsky speculated on *radiation showers* with random results:

> . . if the genes of the germ plasma should be the target of a collision with a cosmic ray or secondary radiation, a mutation of the progeny might ensue; and should many such hits occur, the origin of a new species, most probably incapable of individual or genetic life, but in some cases capable, could be expected. . . . In order for a simultaneous mutation of many characteristics to occur, with a new species as a resultant, a radiation shower of terrestrial or extraterrestrial origin must take place.[22]

Our limited experience with artificially irradiated organisms seems to be heavily (perhaps all) negative. We witness freaks, monsters and genetic disorders. However, there are alternatives to radiation. On the basis of alleged new plants discovered in World War II bomb craters, Velikovsky postulated thermally induced genetic metamorphosis.[23] We also should recognize the possibility of such changes from electromagnetic or chemical assaults.

Thus, according to Velikovsky, we are "led to the belief that evolution is a process initiated in catastrophes."* He continues:

> Numerous catastrophes or bursts of effective radiation must have taken place in the geological past in order to change so radically the living forms on earth, as the record of fossils embedded in lava and sediment bears witness.
>
> The objection to the theory of natural selection, that

---

* In 1985, some 30 years after Velikovsky's suggestion of extraterrestrial-catastrophe-induced evolution, the idea was offered again and tagged as a "spectacular new theory," with no acknowledgment of Velikovsky's priority ["Did Comets Kill the Dinosaurs?" *Time* (May 6, 1985), p.72]. This "new" hypothesis was only one of a continuing line of such "insights" having Velikovsky as their forefather. More and more of Velikovsky's contributions are coming to be accepted -- but always *sans* Velikovsky.

> the developed plan in a new species must appear suddenly or the race would expire . . . is answerable within the framework of catastrophic evolution; however, the purposefulness of animal structures will remain a problem deserving of as much wonder as, for instance, the purposeful behavior of leucocytes in the blood that rush to combat a noxious intruder.[24]

The geological record does suggest that speciations have occurred in the midst of great cataclysms. But perhaps the crucial mechanism could function independently of them, possibly having been produced also on worlds with a less violent history than that of Earth.

Organized Randomness

Velikovsky sees purposefulness in the midst of chaos.

No random process has ever been observed that, on its own, became organized, nor has any organized process, on its own, become more complex, either gradually or suddenly. A random collection of parts, with all the time that has ever passed and all the energy ever available, will always be random.

But perhaps "always" should be mitigated. Ilya Prigogine and Isabelle Stengers, in a brilliant study entitled *Order out of Chaos*, demonstrate that in the non-living world there are fascinating exceptions. They cite, for example, convection currents arising from random molecular motion. But such phenomena are transient; they are not self-sustaining.

Prigogine and Stengers would like to extend this concept to solve a major problem of evolutionary theory: the creation of life from non-life.

> [And] life, far from being outside the natural order, appears as the supreme expression of self-organizing processes that occur.
>
> We are tempted to go so far as to say that once the conditions for self-organization are satisfied, life becomes as predictable as . . . a falling stone. It is a remarkable fact that recently discovered fossil forms of life appear nearly simultaneously with the first rock formations. . . . The early appearance of life is certainly an argument in favor of the idea that life is the result of spontaneous self-

organization that occurs whenever conditions for it permit.[25]

But in searching for this organizing process, they are forced to admit that "we remain far from any quantitative theory."[26] A "simple cell," the basic biological building block, is far from simple and far beyond our present ability to synthesize; even our most sophisticated computers cannot compare to it.

If this most basic hurdle could be cleared, I would yet maintain that complexification of a biological system from random deviations is impossible. A biological whole comes from more than a sum of parts. Regardless of whatever else might be involved in the transmutation of species, energy and intelligence are necessary.

The intelligence must assure that a new organism is complete and functionally coherent at the time of its origin. We have made enough observations by now to realize that all living things are well-designed, with no useless parts, despite occasional speculation to the contrary. The concept of vestigial (useless) organs should be passé.

In the case of radiation, the intelligence could be in the form of selected wavelengths and intensities or, certainly more likely, instructions carried by the organism itself -- as, perhaps, a higher-order or control code somewhat analogous to and controlling the genetic code. One might envisage a universal template restricting the structure of the genetic code, resulting in limited variability and the power to disallow harmful or ineffective combinations of parts. Granted, nonfunctional misfits might emerge and perish, but such phenomena would not negate the suggested control mechanism any more than a deformed offspring disproves the genetic control of normal reproduction.

In Grassé's view, "the genetic code can be read in various ways; it is comparable to the keyboard of a piano from which the performer draws an infinite number of musical pieces. The great problem is to discover the source and the nature of the information which recognizes and animates the genetic segments of DNA."[27]

According to a report in *Discover* magazine, biologists have found that all higher organisms appear to have more DNA (substance of the genetic code) than they need. Nothing is known of the purpose of the excess, but some scientists have been attempting to show that it might help to "turn on" or "turn off" genes that code for specific proteins.[28]

The physical basis for our template, or switch, might thus have been discovered already.

Symbiotic (interdependent) relationships are another matter. We can only marvel at the rapidity of the establishment of these relationships, such as that of the yucca moth and its host. In this case, establishment of the relationship necessarily came during the first generation of the yucca plant -- contemporaneously with the first yucca seeds (from an unknown predecessor). How did the moth come to select this plant to the exclusion of all others? Concurrent genetic changes affecting its structure or physiology and associated behavior? Extinction of all other suitable hosts? The creationist is correct in declaring a sudden emergence, although not under the circumstances he alleges.

Hypothesized here is a speciation phenomenon that is externally induced and internally constrained -- the exact opposite of Darwinist speculation.

Unencumbered by the conceptual problems surrounding Darwinism, we can now entertain a tentative theoretical definition of emergent species, as derived from mass speciation:

> An emergent species is a specific group of organisms (sexual or asexual), of any geographic extent, that is (has been) propagated by a different specific (parent) group of organisms as a result of a specific or specific type of physical disturbance of the sex cells of the parent group.

A fundamental surmise is that a species would give rise to either another single species or to a genus, consisting of multiple species. The latter could result, perhaps, from variations around the mean value of the disturbing force over a geographical range, and/or variations in "response" by different populations of the original species.

Although resulting from a specific cause at a specific point in time, identification of these natural species would not always be conclusive. As Mayr suggested, morphology and interfertility would be the criteria -- perhaps the only criteria for the present. Asexual species could be identified tentatively from morphological characteristics. Reproductive isolation of sexual species -- an *artifact* of speciation, not the cause -- would be a *clue* to species scope -- keeping in mind the possibility of the hybridization of two different species.

Summary

The term "creation" is appropriate for the origin of life. The scope and magnitude of this awakening will never be determined, but there is no objective basis for ascribing all life to a common ancestry. The first step seems to entail an incomprehensible dimension. A mixture of the right chemicals is not sufficient in itself to make this initial leap; we must also account for structure, metabolism, replication. But despite all the requirements or attributes for life that scientists are able to cite, they are still unable to offer a definition for it.

The term "evolution" in a biological context should be dropped; there is no such thing as evolution in the sense generally accepted. Evolution in the accepted sense is implicitly an ongoing process of speciation, a constant modification forced by unrelenting environmental pressures. Such is not -- and has not been -- the situation surrounding speciation. Modifications of the Darwinian kind are limited by the genetic code to variations within the bounds of the species, as with the peppered moth and the fruitfly.

Alfred de Grazia has described the emergence of new life-forms as *hologenesis* -- the creation of new species, structurally and functionally complete.[29] Speciation is not evolutionary. Although recurrent, it has consisted of temporally discrete *historical* changes in biological organisms. These changes were rapid -- from one generation to the next -- occurring only at specific times in the past. Between these times, and into the present since the last upheaval, nature has been virtually static.[30] To reverse a popular adage, "Stability, not change, is the rule of life." But the very beginning, i.e., initial creation, is lost to our view.

Three means of (sexual) species modification are suggested by the evidence:

1. Natural selection: operating on the gene frequencies within populations of a species, allowing the organism to flourish by remaining compatible (or by becoming more fully compatible) with its environment (within limits). Individual members always remain interfertile. This fact can be observed in breeding experiments.
2. Subdivision of a species: multiple internally coordinated changes in the genetic structure producing races, or subspecies, all remaining interfertile. Such subdivision is suggested by natural evidence and ancient tradition to

have occurred once in man (see Chapter XI). It is presumed to have occurred (possibly many times) in other species.

3. Speciation by the transformation of an existing species: one or more species arising from an existing species, with loss of parent-offspring interfertility (the parent species might or might not continue to exist). Such speciation results from internally coordinated changes in chromosomal and/or genetic structure. It is shown by the fossil record to have occurred several times in Earth's history.

Speciation would seem to be greatly accelerated at times of catastrophe. Many forms of life appear to begin or end with the geological divisions examined in previous chapters. Moreover, the Roman poet Ovid (43 B.C.-A.D. c. 17) suggests that some such changes have occurred within man's memory:

> And though fire and water are naturally at enmity, still heat and moisture produce all things, and this inharmonious harmony is fitted to the growth of life. When, therefore, the earth, covered with mud from the recent flood, became heated up by the hot and genial rays of the sun, she brought forth innumerable forms of life; in part she restored the ancient shapes, and in part she created creatures new and strange.[31]

And we find many similar remembrances in folklore from all over the world, most of them associated with catastrophe.

Only one of the means of transformation delineated above is generally recognized: natural selection. It is burdened by the faith that it is responsible for the torturously slow development of all life on this planet.

Science's belief in natural selection as the universal transmutator of life-forms is indeed only faith. Such a role for natural selection is an unfounded extrapolation of a relatively limited process, although I do not intend to minimize the complexity of this process. The protean role assigned to natural selection collapses before an objective analysis of the facts.*

A Goal-Directed Planet ?

The notion of speciation by natural selection is extremely appealing because on the surface it appears to be so logical -- and it seems so elegantly to dispense with the necessity of the supernatural. Yet in all respects this "process" fails miserably.

Unfortunately, what *appears* to be the truth about speciation is perhaps not so appealing. We see in the midst of extreme environmental change the sudden arrival (or birth) of new species that are fundamentally compatible with the new environment. The philosophical explanatory term for such a phenomenon is *teleology*, that is, goal-directed or having purpose. (This term contrasts with *determinism*, or explanations for physical phenomena based on a prior cause.) Teleology is a term that is difficult to avoid in a philosophical examination of much of biological science; it is the oldest problem in the philosophy of biology.[32] And we must consider not only the purposefulness of animal structures and their development (embryology#), but also that of our planet's many interrelationships.

Many biological functions typically are described (perhaps *must* be so described) in terms of their purpose,[34] although scientists strive to avoid the necessity of such explanations.[35] (Recall that the "excess DNA" is presumed to have some purpose.) It is claimed that Charles Darwin's achievement was to show how variation and natural selection can give the *appearance only* of teleology.[36] Hence, natural selection is

---

* Much of what is perceived as "evolutionary adaptation" might actually result from a process of individual development -- growth along a path determined by environment during infancy and childhood. Consider in this regard the pulmonary systems of those who dwell in high altitudes. If an individual is permanently relocated to a lower altitude early enough in life, development is markedly different from what it might have been. Increased lung capacity, for example, doesn't ensue.

"Nothing is more striking in biology than the apparently goal-directed phenomena of embryology and development. And the more we know about the details of development in simple systems and complex ones, the more striking the developmental phenomenon seems. The reason is that not only do whole organs and limbs develop in accordance with a plan, with a goal that is reached even in the face of interference and obstacles, but their component tissues and cells also differentiate and develop, and sometimes even regenerate, in an apparently goal directed way."[33]

mistakenly hailed as having obviated the need for a teleological approach to the explanation of evolution. But even with some kind of "super genetic control" (as proposed in this chapter), when ecological interrelationships are considered, we still seem to be facing purposefulness in nature. Who would describe any biological system or behavior as having no purpose? Reflection upon this and upon the possibility of coordinated change in life-forms and their habitats is cause for much wonderment and awe.

Conclusion

Surely the variety of life around us is not merely the product of mindless Chronos. Time and chance produce only disorder: increased entropy, rust, decay.

No matter how natural the energy source and process responsible for hologenesis, and no matter what the time frame, intelligence was a necessary factor. As suggested above, this intelligence, once begun, is perhaps perpetuated in a natural mechanism capable of directing and coordinating changes in the genetic code. But how far can such an idea be stretched? The existence of *mind* further complicates our quest for understanding.

In an objective pursuit of truth, we should consider all of the available facts -- not just the ones that support a particular viewpoint. We should not be shackled by our subjectivity; we should not abandon logic; we should not create two meanings for a single record, biblical or otherwise, to suit what we wish to believe. We should not ignore kangaroo pockets.

Many facts are in hand, but when all the smoke is penetrated, we find that Truth -- an accurate interpretation of all the facts -- is still somewhere over the horizon.

In any reasoned debate, there should be an unbiased presentation of the facts; hypothesis should be labeled hypothesis. The facts do not include a six-day creation 6,000 years ago, or an ages-long process of speciation by Darwinian forces, or biological complexification by chance.

The issues broached in this book are often emotional ones, and facts that contradict an ideology, secular or religious, will not alone be enough to put them to rest. For the adherents to orthodoxy of both religion and science, rejection of these articles of faith, or acceptance of a new paradigm, will require something akin to a religious conversion.

REFERENCES

1. R. L. Wysong, *Creation-Evolution: The Controversy*, Inquiry Press (East Lansing 1978), p. 34.

2. Lewis Thomas, as quoted in Charles R. Pellegrino and Jesse A. Stoff, *Darwin's Universe: Origins and Crises in the History of Life*, Van Nostrand Reinhold (New York 1983), pp. 188.

3. Hugh Miller, *The Old Red Sandstone* (Boston 1885), p. 48.

4. As described by C. L. Ellenberger, *KRONOS* VII-4 (Summer 1982).

5. *Chemical Engineering News*, October 11, 1976.

6. *Encyclopedia Americana*, "Diatomaceous Earth," Vol. 9, p. 71.

7. Dr. David Herman, *The Prehistoric World of the Dinosaur*, Gallery Books (New York 1988), p. 180.

8. Gerald R. Case, *A Pictorial Guide to Fossils*, Van Nostrand Reinhold (New York 1982), p. 172 (emphasis added).

9. Derek V. Ager, *The Nature of the Stratigraphical Record*, 2nd ed., Macmillan (London 1981), pp. 42-3.

10. Immanuel Velikovsky, *Earth in Upheaval*, Doubleday (Garden City 1955), pp. 28-30.

11. G. G. Simpson, *Tempo and Mode of Evolution*, Haffer Publishing (New York 1944), p. 108.

12. W. R. Thompson, "Introduction" to Charles Darwin's *Origin of Species* (New York 1956).

13. T. H. Huxley, who championed Darwin's theory of evolution through natural selection against clerical and other opposition, chided Darwin in a letter that "you have loaded yourself with an unnecessary difficulty in adopting *Natura non facit saltum* (nature does not make leaps) so unreservedly."

    Tacitly agreeing with Huxley, Pellegrino and Stoff, in *Darwin's Universe: Origins and Crises in the History of Life*, write: "Nature, it seems, does make leaps, sometimes giving rise to new taxa in only a few generations, perhaps even in one" (p. 142).

    As Morton O. Beckner notes in "Darwinism" [*The Encyclopedia of Philosophy*, Vols. 1 & 2, Macmillan (New York 1967), p. 302], the Neo-Platonic notion of Nature as a seamless continuity of forms without gaps "constitutes part of Darwin's cosmology," just as did "uniformitarianism, the belief that nature operates everywhere and always by the same sorts of law." Beckner adds: "This [uniformitarian] view Darwin had imbibed from Lyell's *Principles of Geology*. . . . This particular belief is already a powerful stimulus to look at organic nature as the outcome of a historical process. . . ."

14. *Encyclopaedia Britannica*, 15th ed., "Evolution," Vol. 7, p. 14 (emphasis added).

15. J. W. Klotz, "The Case for Evolution" [as seen by a creationist], *Darwin, Evolution, and Creation*, Concordia (St. Louis 1959), p. 118.

16. Marvin L. Lubenow, "Significant Fossil Discoveries since 1958: Creationism Confirmed," *Creation Research Society Quarterly* XVII-3 (December 1980), p. 148.

17. W. R. Thompson, "Introduction" to Charles Darwin's *Origin of Species* (New York 1956).

18. As cited in Alexander Rosenberg, *The Structure of Biological Science*, Cambridge University Press (Cambridge 1985), pp. 236-7.

19. *Encyclopaedia Britannica*, 15th ed., "Evolution," Vol. 7,

20. Personal correspondence from Roger Wescott (October 1986).

21. Bennison Gray, "The Science of Evolution," *KRONOS* VIII-3 (Spring 1983), pp. 40-1.

22. Velikovsky, op. cit., pp. 256-7.

23. Ibid., p. 254.

24. Ibid., pp. 256-8.

25. Ilya Prigogine and Isabelle Stengers, *Order Out of Chaos*, Bantam (New York 1984), pp. 175-6.

26. Ibid., p. 176.

27. Pierre Grassé, *Evolution of Living Organisms: Evidence for a New Theory of Transformation*, Academic Press (New York 1977), p. 235.

28. James Gorman, "The Case of the Selfish DNA," *Discover* II-6 (June 1981), pp. 32-6.

29. Alfred de Grazia, *Homo Schizo I*, Metron Publications (Princeton 1983), p. 103.

30. Gray, op. cit.

31. Ovid, *Metamorphoses*, I:416-37.

32. Alexander Rosenberg, *The Structure of Biological Science*, Cambridge University Press (Cambridge 1985), p. 38.

34. Ibid., p. 37.

35. Ibid.

36. Ibid., p. 38.

CHAPTER VII

Whence Homo?

The only living species in the genus *Homo* is *Homo sapiens* -- "man the wise." But despite this self-esteeming appellation, our wisdom is limited. There are many things about our very selves that we do not know.

Among the unsolved puzzles crowding our minds is that of our origins. Who -- or what -- were the progenitors of *H. sapiens*? And who came before them?

Anthropologists have unearthed a good deal of interesting information relevant to these questions, but we are tantalized by the apparent voids it contains. Orthodox evolutionary thought demands a continuum of changing ("evolving") organisms leading to man. No such continuum exists. There is no lack of speculation, however, concerning our prehuman ancestors.

Aegyptopithecus is considered to be one of the best transitional forms found in the fossil record, belonging to the Oligocene and Miocene, conventionally dated around 16 to 18 million years B.P. (Before Present) It has characteristics of the prosimians and monkeys, yet its teeth and the form of its eye sockets suggest a connection to the hominoid line, to which man belongs. Many paleoanthropologists believe that *Aegyptopithecus* was on or near the evolutionary line that ultimately split and led to both ape and man. From anthropologists Stein and Rowe: "The primitive nature of the skeletal remains and its great age places [sic] *Aegyptopithecus* in a position to be a common ancestor to all later hominids."[1]

From the upper Miocene (conventionally, from 15 million to 8 million years B.P.) comes *Ramapithecus*. Although not officially categorized as part of the same family as man (Hominidae), *Ramapithecus* is considered to be a strong candidate for the ancestral hominid. In its total structure and form, it resembles the later hominids. For that reason, "many paleoanthropologists consider *Ramapithecus* to be an early member of the Hominidae, a possible ancestor to *Australopithecus* and *Homo*."[2]

The only thing certain about man's possible ancestry is that nothing is certain. Nor will it ever be -- at least, nothing about man's more distant ancestry will ever be. And if, as I have suggested, the supposed temporal sequence of the accepted divisions of the Tertiary is not real, the hypothesized ancestries must be rethought. Either way, the desired continuum will never be discovered since it never existed. Anthropologist Gale Kennedy writes that there is little, if any, evidence of gradual physical change in *Homo* from the Pliocene to the Middle Pleistocene.[3] The creation of man, like the origination of all of the biosphere, was in quantum jumps.* Because of the existence of real genealogical gaps separating distinct types, family trees will continue to be speculative.

Of *Australopithecus*

Man is today the only living erect bipedal (two-footed) creature. The earliest such creature known from the fossil record is alleged to have been *Australopithecus*,[4] conventionally dated from about 3.77 million years to 900,000 years B.P. -- that is, the Late Pliocene to the Early Pleistocene. Hominoid fossils between *Ramapithecus* and *Australopithecus* are extremely sparse.

The *Australopithecus* dental structure was similar to that of *Homo*, although significant differences are noted. The cranial capacity was one-fourth to one-half that of *Homo*; moreover, the cranium was smaller relative to total body size. The foot shows a greater prehensility and

---

* See discussion in Chapter VI.

flexibility than do those of other hominids.[5] As is true for certain other primates, *Australopithecus* had an opposable thumb.[6]

Two major Australopith forms are known: the gracile, found only in Africa, and the robust, having a wider Old World distribution. The robust were larger and more heavily built, sometimes possessing a sagittal crest, a ridge of bone along the midline of the top of the skull to which the temporalis muscle was attached. This crest is seen in the modern gorilla.

The robust form itself consisted of two varieties. The larger (*A. boisei*) is characterized by extremely massive chewing teeth. It is conventionally dated between 2.0 and 1.4 million years B.P. The other (*A. robustus*) is dated between 1.8 million and 0.9 million years B.P. There is no consensus on whether they were separate species.[7]

The smaller and lighter gracile forms (*A. africanus*) appear earlier in the geological record, dated to 3.0 million years B.P. A still earlier and smaller form (*A. afarensis*) is dated to 3.77 million years B.P.[8]

The uncertainty of the evolutionary role of *Australopithecus* is widely recognized. From Stein and Rowe:

> There have been many interpretations of the role of the Australopiths in hominid evolution and the relationship of *Australopithecus* to *Homo*. The views of this relationship have varied from the placement of *Australopithecus*, at least the gracile forms, as a direct ancestor to *Homo* to the view that *Australopithecus* and *Homo* represent very different evolutionary lines, although there may still be an early common ancestor.
>
> The robust Australopiths are certainly too specialized for the role of ancestor to *Homo*. Anyway, we do have direct evidence that the robust Australopiths were contemporary with early *Homo* forms. *A. africanus* is a more acceptable candidate[,] and many paleoanthropologists place *A. africanus* as the direct ancestor to *Homo*. Yet, this gracile form appears somewhat late in the fossil record. However, *A. afarensis* is quite another story.
>
> In the interpretation of hominid evolution put forth by Donald Johanson and Tom White [*Science* 203, 1979], *A. afarensis* is seen as the common ancestor of both the later

> Australopiths and *Homo*. . . In this interpretation *A. afarensis* diverges and gives rise to early *Homo* . . . on the one hand and to the later Australopiths on the other.[9]

So the possible *Australopithecus* ancestry is far from conclusive. *Homo* seems simply to "show up" in the Pleistocene.

Of *Homo Habilis*

Louis Leaky announced in 1964 the discovery of a new hominid species at Olduvai Gorge, in Africa. Christened *Homo habilis*, it has been assigned to a date of 1.85 million years B.P. It was used by Leaky to argue that *Homo* was contemporary with *Australopithecus*. More recent finds also have been placed into this species.

However, when all of the individual specimens are carefully examined, there emerges a mixture of fossils, some of which clearly belong to the genus *Homo* and may be placed with the species *Homo erectus*. The earliest forms, including the initial discovery, fit better into the genus *Australopithecus*.[10,11,12]

The possible reality of a distinct *H. habilis* is an open question.

Of *Homo Erectus*

All in all, a great deal of mystery surrounds the origin of *Homo*. Much of the uncertainty is reflected in the essence of the debate over *H. habilis*. The earliest widely recognized member of the genus *Homo* is not *Homo habilis* but *Homo erectus*.[13,14]

The oldest evidence of *Homo erectus* is dated to the Early Pleistocene. For part of this period, he apparently coexisted with *Australopithecus*.[15,16] Direct evidence of this is said to have been discovered in southern Africa.[17] By the beginning of the Middle Pleistocene, *H. erectus* had spread from the tropics northward into more temperate (European) zones as well as into Asia.[18,19,20]

The cranial capacity of *H. erectus* ranged between 775 and 1,300 cubic centimeters[21]; most specimens fall within the lower range of variation of *H. sapiens*. But according to Stein and Rowe, "the distinctive shape of the *H. erectus* brain case betrays major differences in the development of various parts of the brain housed within it."[22]

Compared to that of *H. sapiens*, the facial skeleton is large relative to the brain case. Behind large and prominent brow ridges, there is a low sloping forehead, the top of the brain case then presenting a long and low profile ending in a prominent angularity in the back.[23] The greatest width of the skull is relatively low, and overall the skull is much thicker. *H. erectus* lacks a chin, and other differences in the jaw and teeth are apparent when these are compared to the jaw and teeth of *H. sapiens.*[24]

The post-cranial skeleton (body) is very similar to that of *H. sapiens*, but many minor differences exist. *H. erectus* is estimated to have averaged five to five and a half feet in height and to have weighed about 117 pounds in adulthood.[25,26,27]

Locations of Remains

Although the remains of *H. erectus* are few and widely separated, they are found almost from one end of the Old World to the other, exhibiting a wide range of dates. Some, however, have not been fully accepted. Sites discovered to date are located in eastern China (two), Java (two), Kenya, Tanzania, Algeria, Germany and Hungary.[28,29]

Transitional Forms

Stein and Rowe continue:

> [M]any specimens [of *H. erectus*] show features that we begin to associate more with *H.* sapiens. In fact, there is no sharp dividing line between *H. erectus* and *H. sapiens*. And we would expect none since one form evolved from the other.[30]

During the Second Interglacial and continuing through the Third (Riss) Glacial, we find a group of fossils which show varying combinations of features, some of which are considered as characterizing *H. sapiens* and others *H. erectus*.[31] But we shall see that the dividing line between species could be obscured for very different reasons.

Of *H. Erectus* Culture

Although the physical remains of *H. erectus* are few, cultural artifacts have been found at many archaeological sites. This culture is known as the *Acheulean*, of the Lower Paleolithic or Lower Old Stone Age. It is characterized by a variety of stone tools: hand axes, hammers, choppers, scrapers, awls and knives. It is also possible that tools of wood, bone, antler, hides, etc., were utilized, but such materials are not often preserved.[32]

H. erectus appears to have used fire, possibly for cooking, possibly for warmth. Fire might also have been used for hunting, to direct the movements of animals. These speculations are based on the discoveries of possible hearths, charred bones, and charred soils where kills were made.[33]

In France has been found alleged evidence of dwellings, thought to have been constructed of rocks and branches.[34]

Archaeologists have excavated sites that are believed to have been *H. erectus* hunting camps.[35] The sites contain bones of now-living species of animals, varying in size from rodent to elephant, as well as bones of extinct mammoths. Stone tools are often found in association with the bones.

Hunting and killing animals as large as elephants and mammoths would require coordinated group efforts. Planning would be necessary, at least to the extent of preparing weapons. Many anthropologists perceive in the activity of cooperative hunting certain "selective pressures" that would have forced the evolution of a larger and larger brain.

> The hunter whose brain was able to remember experiences before reacting to a situation had the edge. Thus natural selection resulted in the enlargement of the frontal areas of the brain, where higher mental activities take place, along with the temporal lobes, where speech is controlled.[36] *

Another "(not-so-) plausible story," yes?

As for speech, we know nothing of *H. erectus'* language or language capability, if any.[37] Most likely, this capability was lacking. Edmund White notes: "The anatomy of *Homo erectus'* vocal tract was probably more like an ape's, smaller and less flexible than a modern human one. He could make some of the sounds we use in speaking but not all, and it appears that he could communicate at only about one tenth the rate of modern man."[38] And while the brain of *Homo* increased

---

* See discussion on natural selection in Chapter III.

in size, no indisputable evidence can be cited to support the concept of a slow, gradual enlargement.

Animal Behavior and Protoculture

Most of the aspects of *H. erectus'* culture, known or presumed, have parallels in the rest of the animal kingdom.

Shelter. Birds build nests that are sometimes quite elaborate. Some are open; others, totally enclosed. Woodpeckers excavate cavities for themselves in tree trunks. Squirrels also build nests, as do gorillas. We do not know what mental or physiological pressures might have urged *H. erectus* to build.

Cooperative hunting. Of wild dogs in eastern Africa, Stein and Rowe write:

> Wild dogs live in packs which contain an average of five to seven adults. . . . These packs range over large areas following the migratory herds which are their prey.
>
> A herd of animals such as the gazelle, the most frequent prey, is stalked. When the herd becomes aware of the pack and begins to flee, the pack begins the chase. The leader chooses the prey, usually the slowest animal, and the rest of the pack follows its lead. If another potential victim crosses the path of a pack member, it will continue to chase the victim chosen by the leader. Members of the pack cut off the prey when it changes its course, thus forming a cooperative unit.[39]

Cooperative hunting thus is not an activity necessarily restricted to creatures of human-level intelligence.

Tools. At the beginning of the rainy season, termite feeding is an important activity for chimpanzees. They might spend up to two hours a day feeding for as long as two weeks.

To reach the termites, the chimp must use a stick. The stick is poked into a hole in a termite mound. Clinging to it when it is withdrawn are termites, which the chimp licks off. The stick usually is about twelve inches long and is manufactured from a

> grass stalk, a twig or a vine. A long stalk, twig or vine section is broken to the right length and stripped of any leaves it might bear.
>
> Young chimps are not interested at first in catching termites. The arts of termite stick making and of termiting must be learned. And both making and using the stick correctly require practice.
>
> Chimps also make sponges from leaves to absorb hard-to-reach water for drinking, and they use other natural objects for tools as well.[40]

Again, such activities obviously do not require human-level intelligence.

> *Protoculture.* Although it includes other things, culture is a body of behavior patterns that are learned and passed from one generation to the next. It is now known that various behavior patterns also are passed down by non-human primates from generation to generation. We sometimes find in animals a "protoculture."
>
> Normally, when macaques are eating, they clean their food by wiping it off with their hands. One day, during a study in Japan, a young female macaque took a sweet potato she had been given and washed it in a stream. Potato washing soon spread to the other members of her play group, then to the mothers of those in the group. Four years later, 80 to 90 percent of the troop were washing their sweet potatoes.[41]

Needless to say, all of these activities are or were accomplished by creatures with no concept of self -- no consciousness. The use of fire (and, of course, ritual) is not expected among these creatures, nor is it seen. We do not expect these animals to use fire because we have never seen such behavior outside human life. And ritual is a product of the human mind.

Caution must be exercised, then, in what we conclude about the mental nature of *H. erectus*. The use of fire could be the only significant aspect of behavior separating him from the rest of the animal kingdom.

Terrestrial Upheaval and Biological Transmutation

I have suggested that major biological changes occurred suddenly in the past and that most of them seem to have come in the midst of terrestrial cataclysm. The available evidence suggests that man was created in the same way.

I have suggested, too, that the eras and periods reflected in the Standard Geological Column actually were *relatively* short times of global upheaval, separated by geologically unrepresented lapses of time of unknown duration. Thus, it is appropriate to refer, for example, to the *Tertiary Disaster* rather than to the conventionally accepted Tertiary Period (which would have preceded the disaster). The biological organization, reorganization and complexification that apparently were initiated during these disturbances is astounding.

Assessment of *Australopithecus*

The oldest proposed date for *Australopithecus* (3.77 million years B.P., based on a single specimen) falls within the Pliocene. The Pliocene-Pleistocene boundary is variously ascribed to different dates, from around 2 million to about 3 million years B.P., depending sometimes on which is the benchmark, climate or ancient animal life. At any rate, the line of demarcation is not precisely defined.

Australopithecus possessed a strange mixture of features, some mirrored today in man and some in apes. Given the manipulative ability inherent in a man-like opposable thumb and the freedom of hand-use associated with bipedalism, the protoculture (?) of the Australopiths was qualitatively different from that of modern chimps. But these creatures undoubtedly lacked the intelligence fully to exploit the potential of these physical attributes, which did Australopiths no good beyond making them dexterous in some activities of "routine existence."

Australopiths became extinct before the Middle Pleistocene[42] -- before, or by the time of, the advent of *H. sapiens*.

Assessment of *Homo Erectus*

Conceived in the Tertiary Disaster by parents uncertain and born to the Lower Pleistocene,[43] *Homo erectus* was mentally superior to his progenitors. His body was almost identical to that of modern man, but his smaller skull housed an inferior brain -- certainly quantitatively inferior. It probably was qualitatively very different. Quite possibly *H. erectus* was a nonconscious being.

In a preceding passage, the frontal (or temporal) lobes of the human brain are cited with regard to speech. Speech aside, the frontal lobes may be the very center and source of our humanness. According to Patricia Goldman-Rakic, an authority on the prefrontal cortex, "If thinking is the process of using information to make decisions, then the frontal lobe is crucial for thinking. Without the frontal lobes, we're at the mercy of our environment. We respond to events without reflection. We are unable to plan for our future. And it is this capacity to plan for the future that distinguishes us from all other species."[44]*

More specifically, Dr. M. Marsel Mesulam, a neurologist with the Division of Neuroscience and Behavioral Neurology at Harvard Medical School, gives a description of patients who have suffered damage to the motor-premotor component of the frontal lobes:

> Some of these patients become puerile, profane, slovenly, facetious, irresponsible, grandiose, and irascible; others lose spontaneity, curiosity, initiative, and develop an apathetic blunting of feeling, drive, mentation, and behavior; others show an erosion of foresight, judgment, and insight, lose the ability to delay gratification and often the capacity for remorse; still others show an impairment of abstract reasoning, creativity, problem solving, and mental flexibility, jump to premature conclusions, and become excessively concrete or stimulus bound.[45]

In summary, Mesulam reports:

> [F]rontal lobe damage interferes with those aspects of our thinking that distinguish us from the rest of the animal kingdom: reasoning, abstract thinking, the organization of behavior over time and space, the attainment of future goals, . . . feelings of personal autonomy and identity, and the ethical and moral components of behavior. . . . [I]t is clear that in order to function as thinking human beings, we must rely on the integrity of our frontal lobes.[46]

Based on his cranial structure, *H. erectus* apparently did not possess the faculties enumerated by Mesulam as distinguishing

---

* But we have just seen that the chimpanzee also is capable of planning for the future, however limited this may be.

humankind "from the rest of the animal kingdom."

Except for the use of fire, all known aspects of *H. erectus'* culture are found today in lesser degrees in nonconscious creatures. But a greater degree of intelligence than that of modern animals might lead to the exploitation of fire despite other mental shortcomings. Such exploitation might require nothing more than a lack of fear and some minimal amount of physical dexterity.

The use of ritual by such a creature of course would be impossible. Ritual could arise only if there existed a mental distinction between self and nature. Neither would there be true language.

H. erectus was modern man's (i.e., *H. sapiens'*) immediate predecessor. The birth of *H. sapiens* apparently was close in time to the beginning of the Middle Pleistocene, and the fossil record suggests that *H. erectus* and *H. sapiens* were contemporaries through much of the Middle Pleistocene, if not to its close.[47] In looking for the cause of the genetic rearrangement that led to the begetting of *H. sapiens* by *H. erectus*, it is interesting to note that "[m]ost paleontologists agree on placing the beginning of the Middle Pleistocene at the time of the last major reversal of the earth's magnetic field."[48] One could be pardoned for wondering what other phenomena might have been associated with this reversal -- and, specifically, whether a magnetic field reversal could account for the genesis of *H. sapiens*.

As to supposed transitional forms, Kennedy writes: "There is, in fact, continuing controversy over whether individuals [from the Second European Glacial] . . . represent *Homo erectus* or *Homo sapiens*."[49] It is recognized that the ranges (variabilities) of certain characteristics of *H. erectus* and *H. sapiens* (e.g., cranial size) overlap.[50] Occasional obscuration of *erectus-sapiens* identities should not be construed as "evolutionary development." Distinct populations continued to exist. The possibility of hybridization must not be discounted, for this could well produce an illusion of evolutionary change.

H. erectus, apparently possessing a new mental advantage, jumped ahead of his progenitors. But *H. sapiens* took a great leap. *H. erectus* was subsequently displaced or else destroyed by the cataclysm that terminated the Middle Pleistocene.*

---

* See discussion in Chapter X.

Of *Homo Sapiens*

Despite many perceived shortcomings, *Homo sapiens* is the crowning glory of Nature's creations, notwithstanding anthropologists' frequent assertions that man is just one more member of the animal kingdom. There is indeed a great gulf between man and beast.

This gulf is denied by many, undoubtedly because of a presumed gradual transition from one level of biological organization to another, culminating in man. Perhaps if the gulf were perceived as having occurred suddenly, the separation would be evaluated objectively for what it is. The *Bridge of Gradualism* does not exist in the fossil record.

It is interesting to speculate on the first memories of humankind and their implication for our history once the great leap from *H. erectus* to *H. sapiens* was made.

The most ancient of the world's geographically extensive myths -- those of the "creation" or the "beginning" -- bear so many similarities to one another that the irresistible conclusion is that conscious man witnessed the great cosmic reorganization described by them, a reorganization that preceded what we have come to call the Golden Age of Man. Exactly how long consciousness might have preceded this beginning, this new age, is not yet known. But memories of an earlier age, preserved in myth, do exist.

An analysis of the similarities and differences of the world's store of creation myths suggests that *H. sapiens* had populated much of the Old World during the time prior to the Golden Age. Assuming that an entire geographically extensive population of *H. sapiens* arose at once from a population of *H. erectus*, this diffusion is to be expected given the geographical distribution of *H. erectus*. (Note, too, that the magnitude of the geographical range of early *H. sapiens,* pre-established by *H. erectus,* would offer no firm clue to the duration of the earlier age.)

All things considered, a fair inference would be that the advent of *H. sapiens* and consciousness occurred "not long before" the beginning of the Golden Age.

I suggest that *H. sapiens'* hominid progenitor, *H. erectus*, possessed no consciousness, no self-awareness, no sense of past or future (the use of rudimentary tools notwithstanding). Consciousness, speech and writing (symbolization) arose with *H. sapiens*. Many other universal aspects of human culture arose quickly on the heels of consciousness, a completely new mental horizon -- *Homo* stripped of instinct and thrust upon the strength of his mind for survival.

Awake and Sing

The path to man was not a gradual process with continuing (and now stratigraphically missing) periods of transition. Rather, it consisted of quite discrete steps culminating in a *great awakening*. Likewise, many of our earliest cultural elements were not a matter of protracted development, but came with the first generations.

One cultural element that came with the very first generation was language. And only within a framework of sudden, complete humanity can we hope to solve the mystery of the origin of language.

THE TAXONOMIC CLASSIFICATION OF *HOMO*

KINGDOM: Animalia

PHYLUM: Chordata

SUBPHYLUM: Vertebrata

CLASS: Mammalia

SUBCLASS: Theria

INFRACLASS: Eutheria

ORDER: Primates

SUBORDER: Anthropoidea

SUPERFAMILY: Hominoidea

FAMILY: Hominidae

SUBFAMILY: Homininae

GENUS: *Homo*

SPECIES: *Homo erectus*

Homo sapiens

REFERENCES

1. P. L. Stein and Bruce M. Rowe, *Physical Anthropology* (New York 1982), p. 337.
2. Ibid., p. 342.
3. G. E. Kennedy, *Paleoanthropology* (New York 1980), p. 349.
4. Harold L. Levin, *The Earth Through Time* (Philadelphia 1983), p. 501.
5. Edmund White, *The First Men* (New York 1973), pp. 12, 14, 15.
6. Kennedy, op. cit., pp. 41, 230.
7. *Encyclopaedia Britannica*, 15th ed., "Australopithecus," Vol. 2, p. 440.
8. Stein and Rowe, op. cit., p. 353.
9. Ibid., pp. 360-1.
10. Ibid., pp. 351-2.
11. Kennedy, op. cit., p. 209.
12. White, op. cit., p. 11.
13. Levin, op. cit., p. 506.
14. White, op. cit., p. 11.
15. William W. Howells, "Homo erectus," *Evolution and the Fossil Record* (Readings from *Scientific American*) (San Francisco 1978), pp. 157-61.
16. Kennedy, op. cit., p. 295.
17. Howells, op. cit., p. 158.
18. Stein and Rowe, op. cit., p. 369.
19. Howells, op. cit., p. 160.
20. Kennedy, op. cit., p. 295.
21. Levin, op. cit., p. 506.
22. Stein and Rowe, op. cit., p. 367.
23. White, op. cit., p. 14.
24. Levin, op. cit., p. 506.
25. Stein and Rowe, op. cit., p. 370.

26. Howells, op. cit., p. 160.

27. Levin, op. cit., p. 506.

28. Stein and Rowe, op. cit., p. 377.

29. Howells, op. cit., pp. 158, 160.

30. Stein and Rowe, op. cit., p. 369.

31. Ibid., p. 378.

32. Levin, op. cit., p. 506.

33. Ibid.

34. White, op. cit., p. 135.

35. Levin, op. cit., p. 507.

36. Stein and Rowe, op. cit., p. 389.

37. Levin, op. cit., p. 507.

38. White, op. cit., p. 18.

39. Stein and Rowe, op. cit., p. 242.

40. Ibid., p. 235.

41. Ibid., p. 253.

42. Ibid., p. 369.

43. Ibid.

44. Richard M. Restak, M.D., *The Mind*, Bantam Books (New York 1988), p. 267.

45. Ibid., p. 268.

46. Ibid.

47. Stein and Rowe, op. cit., p. 370.

48. Kennedy, op. cit., p. 306.

49. Ibid., p. 349.

50. Ibid., p. 315.

CHAPTER VIII

The Origin of Language

Communication is a general term employed in many disciplines. It can be very simply described as meaning that some stimulus or message is transmitted and received. In regard to animal life, it means that information is transmitted from one organism to another, or it perhaps simply involves the recognition of an object, animate or inanimate.

Animals can communicate a variety of messages. These include fear, hunger, sexual receptiveness, status, presence of an intruder, etc. Such communication does not necessarily imply thought. Methods, mechanisms and potentials of communication are functions of neurological complexity,[1] but thought is not necessary for many types of communication.

Communication is a constant process among social animals (primates, for example). The non-human primates are versatile in this respect, exchanging information in four different ways: olfactory, tactile, visual and auditory. As to auditory communication, some apes are numbered among the loudest and most vocal of all mammals. The different sounds and messages communicated by these and other primates are numerous. According to anthropologists Stein and Rowe:

> Sounds can be produced by one animal which direct the

> attention of another toward a specific object, convey quantitative information, specify a particular type of behavior that should be used, or initiate a whole sequence of related behaviors. Primates also have a great ability to inform one another about their moods at particular moments through subtle changes in their vocalizations.[2]

There remains, however, a vast difference between such vocalizations and language. Linguist Mario Pei writes: "Animal cries . . . are characterized by invariability and monotony. Dogs have been barking, cats meowing, lions roaring, and donkeys braying since time immemorial."[3] The same could be said of primates. Even though their vocal communications show many features characteristic of human language, their calls are limited. They do not invent new calls for new situations. [4]

Language Is for Humans

Linguist B. L. Whorf once made the statement, "Speech is the best show man puts on."[5] Indeed, linguistics, the description and analysis of language, is one of the four major branches of anthropology.

Unlike animal cries, human language exhibits infinite variability. The very essence of all language is activity and change. "Even so-called dead languages partake of this changeability," writes Pei, "as evidenced by the ingenious combination devised by the Vatican to express the modern concept of 'motorcycle' in Latin -- *birota ignifero latice incita* ('two-wheeled vehicle driven by fire-bearing juice')."[6]

Human language is open. This means that it has the quality of being able to create new labels for new objects and new concepts. A physicist names and describes a newly-discovered elementary particle. A "primitive" in a remote corner of the world attaches a descriptive name that he understands to a helicopter he sees for the first time and does not understand.

Human language is also discrete. Discrete entities or expressions are represented by words, and these words are arbitrary -- that is to say, the word has no real connection to the thing to which it refers. The potential for sound formation is innate, but the meanings of the elements of a language must be learned.

One of the most important characteristics of language is *displacement*. Displacement is the ability to convey information about things past or future, about things not within sight of the speaker. Because of this, we can learn from the past and plan for the future. We can imagine and create.[7]

Julian Jaynes declares that the chasm between man and the rest of the animal kingdom is awesome. He adds:

> The emotional lives of men and other mammals are indeed marvelously similar. But to focus upon the similarity unduly is to forget that such a chasm exists at all. The intellectual life of man, his culture and history and religion and science [and language], is different from anything else we know of in the universe. That is a fact. It is as if all life evolved to a certain point, and then in ourselves turned at a right angle and simply exploded in a different direction.[8]

And indeed, it did.

Learning To Speak

Languages around the world show a remarkable variability in structure. Outside the phonics typical of most European languages, we find others incorporating strange-sounding clicks and whistles and others with an increased importance of tonal inflections. Some of the more complex languages are found among the more primitive societies.

Yet no matter how complex a language may be, normal children born to its speakers readily learn it, at an equally early age everywhere. The ease of acquisition is an innate ability, an ability receptive to whatever language impinges upon the ear. We are told by linguists:

> It has been shown time and again that a child learns to speak the language of those who bring him up from infancy. In most cases these are his biological parents, especially his mother, but one's first language is acquired from environment and learning, not from physiological inheritance. Adopted infants, whatever their race or physical type, and whatever the language of their actual parents, acquire the language of the adoptive parents who raise them just as if they had been their own children.[9]

"Theories" on the Origin of Language

Mario Pei states: "If there is one thing on which all linguists are fully agreed, it is that the problem of the origin of human speech is still unsolved."[10] He adds, however, that speculation has not been lacking.

Many ancient myths and legends proclaim that language was a gift from the gods. Even as late as the seventeenth century, a Swedish philologist claimed that in the Garden of Eden God spoke Swedish, Adam spoke Danish, and the Serpent spoke French. At a 1934 Turkish linguistic conference, it was argued that Turkish is at the root of all languages, that all words are derived from *günes*, the Turkish word for sun.[11]

Darwin himself offered a proposal: that the beginnings of speech were nothing but mouth pantomime, in which the vocal organs unconsciously attempted to mimic gestures by the hands! Pei called this idea "quasi-scientific."[12] If this notion had not come from an esteemed scientist, Pei perhaps would have given it a more appropriate label.

Most modern theories seem to be variations around a common theme: the gradual transformation of sounds into words.

The "bow-wow theory" maintains that language arose in imitation of sounds occurring in nature, such as the dog being designated bow-wow after his bark. Of course, the same natural noise can have many different human interpretations.[13]

The "pooh-pooh theory" suggests that language first consisted of exclamations of surprise, fear, pain, pleasure, etc. This is sometimes paired with the "yo-he-ho theory," attributing the beginning of language to grunts of physical exertion, and also with the "sing-song theory," holding that language arose from primitive inarticulate chants.[14]

If there is to be found a "consensus statement" on the origin of language, probably one found in *Encyclopaedia Britannica* can serve as well as any other:

> Man's imitation of what is heard, including the sounds of other animals, must be one root of origin. Vocal language must have arisen as a modification of already existing vocal systems. It can easily be postulated that language arises in a species in which auditory control over vocalization is sufficiently developed to permit the individuals to imitate each other's sounds.[15]

Modern linguists, however, are quite pessimistic in their assessment of chances of solving the mystery of the origin of language. According to Pei, "[T]he truly scientific study of the origin of language can properly begin only with the beginning of written-language records."[16] But such records do no better than indicate the former existence of still earlier -- and fully developed -- language. Moreover, no satisfactory theory will ever be forthcoming from today's orthodoxy: Such theories are -- and undoubtedly will continue to be -- directly or indirectly constrained by the artificial limitations of uniformitarian philosophy.

Language and Consciousness

Jaynes has attempted to establish a link between language and consciousness. In doing so, he has suggested that language (and *written* language, at that) *preceded* consciousness. He claims that written language was somehow responsible for the *advent* of consciousness.[17]

I must take a superficially uniformitarian viewpoint in refuting Jaynes. I would argue that consciousness has always been a characteristic of our species. No matter how difficult it might be to define, it is as much a part of our makeup as our ears. Moreover, the Incas, as well as other cultures, never developed writing. Are we to assume that they were nonconscious? This question need hardly be asked.

Logically, both language and consciousness would be prerequisites for the development of written language.

But Jaynes's perception of the relationship of language and consciousness is interesting: ". . . *each new stage of words literally created new perceptions and attentions. . . .*" (italics his).[18] Words did not "create" consciousness, as Jaynes maintains, but surely they do play a role in widening our perspectives -- "stimulating new perceptions" (Jaynes's words) as part of our consciousness.

What is the link between consciousness and language? According to *Encyclopaedia Britannica*: "With the examination of the actual and the probable historical relations between thinking and speaking, it becomes more plausible to say that language emerged not as the means of expressing already formulated judgments, questions, and the like, but as the means of thought itself, and that man's rationality developed together with the development of his capacity for speaking."[19] But it makes no sense to speak of a developing (i.e., evolving) rationality. At what point does rationality begin? At what point does nonlanguage ease across the threshold to become language? In the orthodox scheme we cannot say that these begin when humanness begins; there is no clearly defined or definable threshold for humanness. The link between language and consciousness must be traced within the individual.

Language and latent consciousness possibly codevelop in the infantile human individual. Consciousness is, and must be, part of our genetic makeup. In the individual, language and thought must serve to expand each other. (These suggestions are clouded, however, by consideration of children deaf and/or blind from birth.)

Man the Wise

In the preceding chapter, I maintained that *Homo sapiens* was born from another species, *Homo erectus*, and that this birth was not a gradual, evolutionary process but was a sudden one. At some point in the history of the genus *Homo*, a given generation of the elder species produced the first generation of the younger. A nonconscious being gave rise to a new creature with a conscious mind (that is, with latent consciousness); the new creature was unlike any other in the history of the world.

That creature's mind was every bit as developed and capable as the mind of man today. At birth, this first generation was comparable to any group of newborn infants in a modern hospital.

The difference in the beginning was the absence of culture to provide material for "programming" the newly created human. But the mind was there -- a mind that would lead to the very first aspect of human culture: language.

Let There Be Speech

Some animal vocalizations are instinctive, some learned. In either case, they have no abstract, symbolic meaning.

When a human child learns the language of its parents, it is not somehow provided with the explicit definitions of words. Rather, through various means of input and feedback, it comes to associate sounds with objects, actions and concepts. The child *symbolizes*. Moreover, as its speech develops, it does not merely mimic; it creates. (Witnessing this in a two-year old is truly astounding!)

We have noted that the sounds of speech are completely arbitrary insofar as their meaning is concerned. To be learned as a language, the only requirement they must meet is that they be consistent. This being the case, the sounds need not be recognized by their user as words, as symbolic representations. They would come to be interpreted and conceptualized as such by a newly-born, learning, intelligent hearer possessing latent consciousness.

Thus, the first speech: not random sounds that gradually developed meaning over centuries or millennia of time to be finally recognized as words, but consistent (non-language) sounds unconsciously transformed into abstract representation (words) by an infantile conscious mind. The young *H. sapiens* heard the same sounds as his *H. erectus* parents, but for him the basis of meaning was of a totally different nature. The physiology and character of the enunciation would likewise have been different.

An initially limited vocabulary rapidly grew with peer contact and (immediately possible) linguistic innovations. Progress through time would have been very rapid, and the development of different (i.e., geographically separated) linguistic structures was cast at the outset.

Awareness did expand with language; one drew from the other; and the young *H. sapiens* could communicate much of his own private, internal awareness with words -- with sounds that had no meaning except within the context of a shared internal awareness.

An interesting parallel to this surmise is found in the research of Dr. Derek Bickerton, as cited by Richard Restak. Bickerton is a linguist who studies languages called *Creoles*, which are created when established languages are useless. According to Bickerton:

> . . . [And] when slaves were taken away from different areas of Africa, they spoke different languages. . . . [T]hey were bought on the slave market by owners of different plantations, who also spoke different languages [from those of the slaves]. People can't stop communicating with each other, and in that process, you develop a lot of languages.
>
> Many slaves . . fled to form their own communities. It was in these communities that they evolved pidgin speech, an improvised language that had a sparse vocabulary and no real grammar. Pidgin made communication possible among people who had no common tongue. Thus, their children also lacked a true

> language. By spontaneously bringing grammar to their parents' pidgin, . . . the children created a completely new language in one generation. This language is a Creole.
>
> The adults who make pidgin are not able to provide it with any structure. They're past the critical age at which syntax develops. The children, however, are not. Syntax develops in them just as naturally as any other physical part of their bodies. It's natural, it's automatic, it's instinctive, and you can't stop them from doing it. I think the only explanation you can have for the way syntax works is that somehow, this is built into the hard wiring of the neural circuits of the brain. . . .
>
> Pidgin is the first stage in [an effort] to communicate with each other. Creoles are an order of magnitude different. They are full languages, rich in syntax even if limited in vocabulary.[20]

And such must have been the case with the first languages.

A Note on Writing

The origin of the facility for speech was strictly a biological/physiological phenomenon, the mental and structural requirements having emerged together and at once. Writing, a cultural innovation, was initiated afterward, whether in the same generation or centuries later.

The very first step in this process undoubtedly would have been the recording of a simple symbol recognized by a group as representing/depicting something with which they were familiar. Spoken and written language would develop rapidly thereafter and expand hand-in-hand.

And perhaps this first symbol had its roots in a celestial apparition visible to all. (Recall the Turkish "günes.") We close here with that thought directing our attention to the heavens.

REFERENCES

1. P. L. Stein and B. M. Rowe, *Physical Anthropology* (New York 1982), p. 246.
2. Ibid., p. 247.
3. Mario Pei, *The Story of Language* (Philadelphia 1965), p. 23.
4. Stein and Rowe, op. cit., p. 247.
5. B. L. Whorf in J. B. Carrol (ed.), *Language, Thought, and Reality: Selected Writings of Benjamin Whorf* (Cambridge 1956), p. 249.
6. Pei, op. cit., p. 23.
7. Stein and Rowe, op. cit., p. 247.
8. Julian Jaynes, *The Origin of Consciousness in the Breakdown of the Bicameral Mind* (Boston 1976), p. 9.
9. *Encyclopaedia Britannica*, 15th ed., "Language," Vol. 10, p. 649.
10. Pei, op. cit., p. 21.
11. Ibid.
12. Ibid.
13. Ibid., p. 22
14. Ibid.
15. *Encyclopaedia Britannica*, 15th ed., "Hominidae," Vol. 8, p. 1,027.
16. Pei, op. cit., p. 27.
17. Jaynes, op. cit., p. 66.
18. Ibid., p. 132.
19. *Encyclopaedia Britannica*, 15th ed., "Language," Vol. 10, p. 644.
20. Richard M. Restak, *The Mind*, Bantam Books (New York 1988).

CHAPTER IX

The Mysterious Origin of the Moon -- and the Non-Moon of Genesis

The moon is one of the most beautiful objects in our sky and, next to the sun, by far the most conspicuous. It comes and goes bit by bit every month. We look at it. We study it. We still write poetry about it even though some of its mystery and mystique have faded since we visited it.

It is our constant companion. Has it not always been so? Scientists tell us that it preceded our species by millions of years.

There are three major competing theories of the moon's origin. One holds that it and the earth condensed out of the same cloud of stellar dust at the same time long, long ago. A number of factors, some of which are unadmittedly fundamental, cannot be explained by this notion, but it has a tenacious grip on many minds.

Another proposes that it came from the distant reaches of the solar system and was captured by Earth more than 4 billion years ago.

The third is the concept of cleavage, or terrestrial fission -- the idea that the moon, by an uncertain mechanism, was ejected from the earth. This event usually is ascribed to an ancient prehistoric period, but unorthodox proposals also are heard for an era after the advent of man. After falling from popularity, the theory of terrestrial fission is experiencing a revival.[1]

Now orthodox cosmogonies, like orthodox geology, adhere religiously to uniformitarianism, with no regard for ancient records that indicate a different history for Earth and its moon.

Or do any ancient stories indicate an other-than uniformitarian history for the moon? Whose culture can remember when the moon was *not* with us? The memory of its ancient presence seems to be preserved in some of our oldest literature, the Book of Genesis: "And God made the two great lights; the greater light to rule the day, and the lesser light to rule the night. . . ." (Genesis 1:16). Regardless of whether they classify the first chapter of Genesis as history or as poetry, believers and skeptics alike have always agreed on how to interpret the terms of this passage. What could be more obvious than the "greater light" of the sun to "rule the day" and the "lesser light" of the moon to "rule the night," even though the sun and moon are not explicitly identified? Within our experience, this seems to be the only possible interpretation.

Moonless Nights

In March 1973, however, I suggested that "lesser light" might be a reference to a genuine source of light -- a light-emitting body unknown to modern eyes -- rather than to the moon.[2] This same idea was later put forth in a claim that there is undeniable evidence that the concept of a NIGHT-sun as well as a DAY-sun existed in ancient Babylonian thought; according to authors L. M. Greenberg and W. B. Sizemore, these "suns" were viewed, respectively, as the lesser and greater chief lights of heaven.[3]

Could the moon actually have been preceded by, and now be confused with, an object no longer known or recognized?

In May 1973 Velikovsky published a short article citing several references to an alleged ancient era in man's memory when there was no moon. In summary, he states:

> The traditions of diverse people offer corroborative testimony to the effect that in a very early age, but still in the memory of mankind, no moon accompanied the earth. Since human beings already peopled the earth, it is improbable that the moon sprang from it; there must have existed a solid lithosphere, not a liquid earth. Thus it is more probable that the moon was captured by the earth.[4]

The truth of this allegation might only imply that the Genesis account was recorded after a lunar capture. But a late appearance of the moon would explain the lack of a reference to months in the first chapter of Genesis even though days, seasons and years are mentioned. Months are correlated, of course, with the moon's orbital period.

The first biblical reference to months is in relation to the chronology of the Deluge. This allusion implies that the moon was no stranger to man at that time. Between the first chapter of Genesis and the Deluge, however, there is no allusion to the introduction of the moon. We must look elsewhere if we are to find a more definite contrast between the Lesser Light and the moon.

A Diminished Light

The sought-after connection resides in various traditions -- a commingling of recollections contrasting two quite dissimilar objects -- suggesting that the nature of "the moon" was not always the same as we know it to be.

From the Bomitaba in Africa:

> Once upon a time there were two suns, the one we have and the moon. It was very tiresome for mankind, which being constantly in heat and light could not rest comfortably. One day one of the suns suggested to the other that they should both bathe, and pretended to jump in[to] a river; the other threw itself in and was quenched. Since that time there is only one sun, and though the moon lights men it no longer warms them.[5]

A somewhat similar myth from tribes in Australia states that "in the beginning the Sun never set, but as human beings were weary of perpetual day (that is, of not being able to sleep) the creating deity at last ordered the Sun to set."[6]

The Luyia of Kenya say that "God created the Moon first and then the Sun. *In the beginning the Moon was bigger and brighter*, and the envious Sun attacked his elder brother. They wrestled. . . and the Moon was thrown in the mud and dirt splashed over him so that he was not so bright. God intervened . . . [and said] that the Sun would be brighter henceforth and shine during the day. . . . The Moon would only shine at night."[7]*

And from the Hebrews:

> "Thou didst create the heaven and the earth, the heaven exceeding the earth . . . and now thou hast created the sun and the moon, and it is becoming that one of them should be greater than the other" [said the moon]. Then spake God to the moon: "I know well, thou wouldst have me make Thee greater than the sun. As a punishment I decree that thou mayest keep but one-sixtieth of thy light." The moon made supplication: "Shall I be punished so severely for having spoken a single word?" God relented: "In the future world I will restore they light, so that thy light *may again be as the light of the sun.*"[8]

Finally, the Navajos have a myth that says that since time was assigned, there have been twelve moons (one each month of the year), whereas before there was only one moon.[9] The Navajo myth suggests a celestial object that was constant in the heavens, an object that did not experience each year twelve deaths and rebirths as the moon.

A Justifiable Conclusion

We thus may visualize a magnificent luminous sphere, in addition to the sun, as having been at one time a familiar presence in Earth's sky. Long since forgotten, it left no visible evidence of its former status, no physical record.

And then the moon appeared, making obscure even the written record.

Our solar system is not what it once was. It is not difficult to see in these various traditions veiled remembrances of the change that has been suggested: the passing of a large celestial source of light and the introduction of a smaller and merely reflective, dimmer entity. The

---

* It is perhaps significant that the "moon" was regarded as the *elder* brother, considering the assertions made in the next chapter.

moon did indeed follow a more splendid predecessor* and was captured by our planet within man's memory.

What, exactly, might that predecessor have been? And where is it today? To find out, we must look to "The Beginning."

A digression is warranted, however: We should first examine a recent version of the aforementioned terrestrial-fission notion of the moon's origin.

Catastrophism: Acceptable and Unacceptable

In the November 1989 issue of *Natural History* there appeared one of the most bizarre, self-contradictory articles that anybody could wish ever to read. Its title: "Birth of the Moon." It proposes that the moon erupted from Earth as the result of a stupendous impact of extraterrestrial origin.

After setting the standard uniformitarian stage with the prescribed 4.5 billion-year age of the solar system, William Hartmann informs the world that:

> Otto Schmidt, a Soviet planetologist, argued that if each planet had formed separately from a single blob, their motions would be more random; some planets would travel in highly elliptical orbits in different directions. They would spin in the same direction, and their axes would be randomly tilted. As Schmidt pointed out, the solar system is, instead, quite orderly: with few exceptions the planets all travel in concentric, nearly circular orbits in the plane of the sun's equator; revolve in the same prograde direction . . . and spin on their axes in the same prograde sense, with little tilt. . . .[11]

I had to do an immediate double-take when I first read the preceding passage. Surely it was a printers error that switched the *is system* for the *would-be system*. But considering what follows, that

---

* Dwardu Cardona contends that because of this entity's (former) magnificence and proximity to Earth, people actually regarded *it* as the Greater Light and the *sun* as the Lesser Light.[10]

certainly was not the case. Further into the article, Hartmann states:

> Although the process of planet formation was more or less orderly and gradual, occasional big impacts may have superimposed some variety on an otherwise dull solar system. Evidence of chaos during the formation of the planets is everywhere: Mercury, which is little more than an iron core with a thin rind of rock, may have *experienced an impact large enough to blast away most of its mantle*; Venus, which *spins slowly backward*, may have been struck a glancing blow; both the earth and Pluto have an unusually large moon for their size; Uranus . . . *spins on its side*; and Neptune, which seems to have a disturbed satellite system, may have had *a close encounter with a large body.*[12]

To top things off, he writes that "*the moon was born in one of the most violent catastrophes in solar system history.*"[13]

These passages evoke a burlesque vision of one actor playing all three roles of the proverbial blind men feeling and describing the elephant. How many solar systems is Hartmann describing?

A reader wonders if Hartmann might be trying to help Velikovsky gain a foothold in orthodox astronomy. The whole thrust of the article is a catastrophic origin for the moon. But for the Velikovsky apologist, hope turns to dismay as he continues to read:

> The resistance to the catastrophic nature of our theory was interesting because of its sociological roots. First, all of science has been affected by the principle of uniformitarianism promoted by James Hutton and Charles Lyell: that the earth's complex geology was created by the same processes that are changing the planet today, acting over millions of years. Similarly, Charles Darwin's theory of evolution proposed that life could undergo immense transformations -- given enough time. Those ideas, which saw nature as the result of cumulative, gradual change, triumphed over the older, catastrophic theories, which supposed mountains and species were created by a few sudden and dramatic events. The idea that the moon was

created in one big spectacular splash, instead of a long series of modest events, had a bad ring to it because it harkened back to the era of catastrophist ideas. Furthermore, planetary scientists had been embarrassed by the controversy over the books of Immanuel Velikovsky, a Russian-born physician. In *Worlds in Collision*, published in 1950, Velikovsky postulated that major collisions [Velikovsky mentioned *collisions* only in the *title* of his book] and near misses between the planets had occurred in the last few thousand years and could explain ancient legends such as the Great Flood, the parting of the Red Sea [Velikovsky made no reference to "the parting of the *Red Sea*"], and Joshua's causing the sun to stand still (the earth briefly stopped turning in one of Velikovsky's near misses). Astronomers could prove [?] that the solar system had not been so chaotic in the last few thousand years, but in spite of hundreds of lectures arguing against such catastrophism, Velikovsky's books became best sellers and the misconceptions resulting from his pseudoscience spread. *Our new theory, which attributed the creation of a major feature of the solar system to a giant catastrophe, was therefore not welcome.*[14]

What a pity! And Hartmann's is such an original, self-consistent, convincing view of our solar system! He could summarize his position as follows:

1. It's obvious that the solar system has always been uniform and peaceful.
2. It's obvious that the solar system has suffered great catastrophes because of its many non-uniformities.
3. Velikovsky's catastrophes should be ignored because they are pseudoscience.
4. My catastrophe should be accepted because . . . ?
   Because . .
   Oh, yes -- because it is a *uniformitarian catastrophe* and I am anti-Velikovsky!

As has happened repeatedly since Velikovsky's first publication, his hard documentation is belittled and his claims misrepresented as pseudoscience while pure speculation is lauded as science -- and in Hartmann's case, with an implied measure of self-praise to boot. Such perversity makes for wonder: How does one discern between good catastrophism and bad? And why such a fear of giving Velikovsky any credit?

The answer may lie partially in the timing of Velikovsky's work. He came at a time when Neo-Darwinism felt it had won the day from the last die-hard scientific opponents yet, perhaps, when it was not fully recovered from the violence of popular feelings stirred up during the Scopes trial in 1925. Indeed, Velikovsky's *Worlds in Collision* was published in 1950; the Tennessee anti-evolution law that provoked the Scopes case was not repealed until 1967. How could the uniformitarians, who had such a big stake in Neo-Darwinism, feel safe in 1950? Perhaps the violence of their attack on Velikovsky is to be attributed to fear. To this day, biblical fundamentalists are using the work of Velikovsky as core material for their guerrilla war on uniformitarians.

Among other elements of his hypotheses for which Velikovsky was castigated was the recentness of the upheavals he described. This remains today a major source of complaint against Velikovsky, but because of his "original sin" in undermining uniformitarians just when they felt they had seized the day, nothing that Velikovsky ever proposed is acceptable -- that is, not unless it can appear as a "new insight" by a "member of orthodoxy" without any manifest damage to the uniformitarian framework -- and (praise Darwin and pass the fossils) without acknowledging any contribution whatsoever by Velikovsky.

Despite heated opposition, Velikovsky's work has withstood every uniformitarian challenge to be marshaled against it. (See, for instance, *Scientists Confront Scientists Who Confront Velikovsky*, KRONOS Press, ed. by L. M. Greenberg.)

Today, much of the problem results from complete ignorance of Velikovsky's works (although in many instances the same can be said of the initial outbursts). A generation of scientists has been convinced by hearsay that Velikovsky was a crackpot.

In his *Natural History* article, Hartmann writes: "The Lunar samples . . . indicated that whatever the moon's origin, its materials had once been heated to high temperatures by some violent process."[15] Such findings but strengthen Velikovsky's claims upon posterity. Based on his general theory, and *before any evidence was in hand to suggest it*, Velikovsky made the claim that the moon had been heated by catastrophe -- to the point of its surface becoming molten. And this claim was published at least as early as 1969 (*New York Times*, July 21, 1969, and later in *Pensée* II-2, May 1972).[16] Hartmann probably never knew of the Velikovsky prediction, else surely he would have given credit where credit was due. He pleads eagerly enough for his own credit.*

It is becoming more and more difficult to reconcile the treatment of Velikovsky with currently acceptable speculations and perspectives within orthodoxy. While ridicule is heaped on the notion of planetary encounters (see Chapter I, where Harold Levin's 1983 attack on Velikovskian catastrophism is cited), we also read a 1988 assertion that "Halley's comet *proves* that such bodies may collide with Earth."[17]

Proof -- without even the benefit of a near miss. Hard evidence, so abundant yet so often denigrated, is welcome to people who would like to see Velikovsky vindicated, but catastrophists can live without such "proof" as this 1988 allusion to Halley's comet.

The uniformitarian looks more than ever like somebody who likes to have his cake and eat it, too. The Plausible Story, of course, is an ever-renewable confection.

---

* Hartmann's article should stand as a monumental embarrassment to his colleagues and to the editors of *Natural History*. Although I have no documented support at hand, I nevertheless suspect that these same editors have rejected numerous pro-Velikovsky articles submitted to them for publication -- articles that no doubt were professionally written and made sense.

REFERENCES

1. Encyclopaedia Britannica, 15th ed., Vol. 12, p. 429.
2. J. E. Strickling, "The Birth of the Gods," *Bible-Science Newsletter* XI-3 (March 1973).
3. L. M. Greenberg and W. B. Sizemore, "Saturn and Genesis," *KRONOS* I-3 (Fall 1975).
4. Immanuel Velikovsky, "Earth Without a Moon," *Pensée* III-1 (Winter 1973), p. 25.
5. *Larousse Encyclopedia of Mythology*, 2nd ed. (London 1968), p. 482.
6. Ibid., p. 464.
7. E. G. Parrinder, *African Mythology*, Hamlyn (London 1967), p. 71 (emphasis added).
8. Louis Ginzberg, *Legends of the Jews*, Jewish Publication Society (Philadelphia 1913) (emphasis added).
9. C. A. Burland, *North American Indian Mythology*, Hamlyn (London 1968), p. 93.
10. Dwardu Cardona to J. E. Strickling, personal correspondence (March 1985).
11. William Hartmann, "Birth of the Moon," *Natural History* (November 1989), p. 68.
12. Ibid., p. 74 (emphasis added).
13. Ibid., p. 76 (emphasis added).
14. Ibid., p. 74 (emphasis added).
15. Ibid., p. 73.
16. Thomas Ferté, "A Record of Success," *Pensée* II-2 (May 1972). pp. 13-14.
17. Dr. David Norman, *The Prehistoric World of the Dinosaur*. Galley Books (New York 1988), p. 152 (emphasis added).

CHAPTER X
In the Beginning

In the Book of Genesis we read, "In the beginning God created the heaven and the earth" (Genesis 1:1). Thus begins a narrative that purportedly describes the origin of all things, inanimate and animate alike. It is perceived by many as the singular, accurate account of the beginning of time, matter, and life and of the laws by which all things are governed.

For some, the Genesis passage is merely poetry -- an aesthetically pleasing fabrication. For others, it is the foundation of a philosophy of science, reflecting a brief prehistoric period before physical law took effect.

Summarily excluded from the latter category is a multitude of sometimes colorful creation myths from all quarters of the world. Here we find, often in fanciful terms, detailed descriptions of the origins of Earth and all its inhabitants.

How did these ancient tales come to be? Many of the faithful claim that the Creation was supernaturally revealed to the Prophet Moses, who was responsible for its recording in Genesis; man obviously could not witness his own creation or that of whatever preceded him. Accounts other than that in Genesis are less controversial. Almost everyone agrees on their "mythopoeic status."

Far too often, however, we lightly dismiss in this manner the observational records and memories bequeathed to us by the ancients. Perhaps this is another instance of a preconceived opinion standing in the way of understanding and enlightenment. If we pause to listen closely, we might learn more about "the beginning." And perhaps the Genesis record itself has more to tell than we ordinarily imagine.

The Golden Age

As is true of creation myths, legends of a Golden Age in the earth's distant past abound in the memories of man. This age apparently was a time when ideal natural living conditions prevailed, conditions unequaled during any other period of world history. Although millennia have elapsed since the end of that age, its melancholic memory persists in our hopes and dreams, continually surfacing in our art and literature.

Various scholars have attempted to winnow the stories in search of a composite vignette of that ancient age. One such is Roger W. Wescott. Based on statements contained in the legends and on inferences drawn from them, Wescott has characterized a world that makes us envy our ancestors, a world untrammeled by the grief and stress of ordinary life today, whether industrialized or primitive.[i] It was, indeed, like a Garden of Eden.

The climate, according to Wescott, was warm and moist, and vegetation was lush. Seasons were nonexistent because Earth's axis had no tilt, i.e., it was perpendicular to the orbit. Without seasons, many of the annual variations and/or cycles that characterize plant and animal life today also would be nonexistent. Wescott maintains, in addition, that the topography of Earth was typified in smooth and rolling hills, without the sharp angularities and craggy peaks so familiar today. (It is unlikely that this was true everywhere, however, given the long, violent history of our planet before the advent of man.)

All animals and people were foragers. Carnivores lived not by predation, but by scavenging. Competition was minimal, and there were no ranked hierarchies or defended territories.

Women did not menstruate, and childbirth eventuated without pain. Disease was rare, and life spans were exceedingly long.

Social organization was uncoercive. Artifacts (such as clothing), institutions (such as marriage), and ideologies (such as religion) were unknown. Relations between individuals were probably pleasure-oriented rather than moral (avoidance of harm). There was no war.

Although this (abbreviated) assessment is essentially compatible with the Genesis picture of the Garden of Eden, Wescott's assertions are not reconcilable with the record of pre-Deluge patriarchs in Genesis 4. There we find referenced a murder, animal husbandry, agriculture, a city (implying trades), metalworking and marriage. Moreover, the Patriarchs believed that the reason behind the Flood was widespread strife (Genesis 6:5-7).

An unavoidable question, then, is which biblical period -- Garden of Eden or Pre-Deluge Patriarchs -- equates to the world's store of legends of the Golden Age? The answer would seem to be the former if conditions are considered, but the latter based on chronology (Golden Age terminated by the Flood). Based on the biblical chronology [Garden of Eden, Golden Age (?), Flood], the time summarized in much of Wescott's description thus appears to predate the Patriarchal Age.

Other accounts than that of Genesis may illuminate this time-before-the-Flood (whatever its scope) when there reigned an all-but-forgotten cosmological and terrestrial setting that almost defies comprehension.

Monarch of the Golden Age

The sky of that age was a thing of wonder; the picture that can be extracted from the message of myth is startling. Here we must draw from the researches of other scholars besides Wescott. Probably the most prolific in this regard have been David N. Talbott and Dwardu Cardona.

As with so much of modern catastrophist research, the foundation for Talbott and Cardona's studies was laid by Immanuel Velikovsky. After presenting evidence that planetary orbits have undergone significant change in historical times,[2] Velikovsky suggested that the earth might once have been in orbit around Saturn before being captured by the sun.[3]

Talbott, Cardona and others pursued this clue through the world's mythological records, reaching a conclusion that staggers the imagination. This conclusion is precarious in that it has not yet been supported by known principles of physics and in that no comparable example has ever been observed. Yet the myths seem to be universally consistent in what they describe, and it has been suggested that there may be astrophysical principles with which we are not yet acquainted.

However, some researchers deny totally that what Talbott and Cardona describe could ever have happened.

In the long run, the difficulty must be resolved, but the mythological message continues to tantalize. The essence of the message? Cardona maintains that "the planetary system that our ancient forefathers inadvertently described was an entirely different, and quite bizarre, celestial arrangement. What we see portrayed in the sky today pales to insignificance when compared to what the ancients claim *they* saw."[4]

What did the ancients see? We find a bold account in Talbott's interpretation of the records.[5]

During the Golden Age, Talbott writes, Saturn dominated the heavens. Saturn -- not the sun -- was the Sun-God. Saturn was far more spectacular, and it fired man's imagination in a way that the sun could never have done. Indeed, Saturn was known to the ancients as the "sun-planet" or the "sun-star."

Saturn was not called a sun because of its being brighter than the sun, which it could not have outshone. Rather, in its frame of a darkened sky it had an incomparably greater impact on the mind and culture of man than did the sun. It appeared as an immense and visually overwhelming orb in the sky, an unimaginably spectacular display of lights.

Associated with, and forming part of, the phenomenon is the global tradition of the Cosmic Mountain. Hebrew legend speaks of Mt. Zion; the Greek, of Mt. Olympus. The Babylonians called it the Mountain of the World. The Hindus remember it as Meru, the Chinese as Kwen-lun, and the "original Mexicans" as Colhuacan. The mountain is said to have appeared as an entity connecting the heavenly orb with the earth as if to support the Saturn-sun -- a pillar of light composed of Saturnian debris, a river of fire and light joining two very different worlds.

To present such an apparition, Saturn obviously would have to have been much closer to Earth than it is today; it now appears as a mere pinpoint of light, unrecognized by the casual star-gazer. Moreover, the records are unanimous in declaring that it rested in the north, occupying the celestial pole; it was the center of cosmic revolution -- just as the pole star is today. And herein is the apparent physical problem: The implication is that Saturn and Earth shared the same axis of rotation.

The column of light sometimes appeared to be like an elongated pyramid in shape. After the Flood, this image was memorialized in the Egyptian and New World pyramids, as well as in similar if less impressive architectural forms in other parts of the world.

The column also was perceived as a vine and/or tree, the World Tree. We memorialize it today in the Christmas tree, with its (Saturnian) star at the top. Therein we also encounter the origin of Santa Claus at the North Pole -- the bearded figure recognized the world over (much as Saturn was) as a culture hero, a bearer of gifts to mankind.

Philosopher Lynn Rose would have the earth in a normal orbit around Saturn, with the same face always turned toward it, as the moon now keeps the same face always toward Earth. This, too, would result in the huge orb appearing to be fixed in the sky. Although such an orbit cannot account for the mythological allusions to the polar configuration, it does not need an appeal to unknown principles of physics. Moreover, Rose suggests that the polar (axis) association arose after the collapse of the system because the pole star became the only apparently stationary object in the sky.[6]

The disruption of the Saturn-Earth system was followed by the Great Deluge,[7] which was probably the most traumatic catastrophe in the history of mankind -- not only the Flood itself, but the loss of the Saturnian presence and a utopia that would be sought ever after. To this day the cultures of the world are riddled with memories of Saturn's various manifestations, including our fairy tales, our nursery rhymes, and some of our most basic observances.

We find these memories in the story of Jack and the Beanstalk, with the Golden Egg at its peak in the clouds of heaven. Incorporating the same roots is the story of Humpty Dumpty, recalling the helpless feeling of mankind at Saturn's demise. And every week we rest on Saturn's Day.

The reader must turn to Velikovsky for more extensive details of Saturn's abdication.[8]

The Earliest Memories of Man

Some of the earliest recollections of man are the stories of the "creation." But how can man remember his own origins? Certainly he cannot. What meaning, therefore, do the stories have? Close listening to the tales may suggest one.

Egypt. In the beginning the universe was filled with a primordial ocean called Nun. Nun had no surface; it completely filled the universe and was likened to a cosmic egg. The waters of Nun were motionless.

The first event in the creative process was the emergence of Atum from the chaotic wastes of Nun. Atum created himself by an effect of Will, or else he was the child of Nun. Finding nowhere to stand, he created a hill upon which to rest. His emergence was accompanied by the coming of light to disperse the chaotic darkness of Nun.

Atum became identified as Ra, or Ra-Atum, creator of the world. It was on the command of Ra that Nut (the sky) was separated from Geb (the earth), thus forming the world.[9]

Greece. Chaos was surrounded by night, which formed the enveloping cover under which cosmic matter was slowly organized. This finally assumed the shape of an egg, of which Night formed the shell.

In the center of this gigantic egg, whose upper section formed the vault of the sky and whose lower section was the earth, was born the first being: Phanes, the light. It was Phanes who, by union with Night, created Heaven and Earth.

On her part Gaea (the earth) first bore Uranus, the sky, crowned with stars. Then she created the high mountains and the sea.[10]

India. One version of the creation relates that the golden cosmic egg was floating on the waters for a thousand years. At the end of this period, the egg burst open to reveal the Lord of the Universe.

Later myths refer in some way to Brahma as creator. One relates that the lord of the universe brooded over the cosmic egg as it lay on the surface of the ocean for a thousand years. As he lay there in self-communion, a lotus, bright as a thousand suns, rose from his navel and spread until it seemed as if it could contain the whole world. From this lotus sprang Brahma, self-created.

Another version is given in the Laws of Manu. Manu wished to create all living things from his own body. He first created the waters and then threw a seed into them. From the seed grew a golden egg, bright as the sun. The self-existent spirit developed within the egg as Brahma.[11]

China. Chaos was like a hen's egg. At that time neither Earth nor

Heaven existed. From this egg, Pan-ku was born. The parts of the egg separated. The heavy elements formed the earth, and the light, pure ones, the sky. For a period of 18,000 years, the distance between earth and sky increased by ten feet daily, and Pan-ku grew at the same rate, so that his body always filled the space between the two.

Chaos had to end before the world could begin. Similarly, the concept of a pillar (instead of a body) keeping apart or joining the two parts of the world is found in Chinese cosmography. Elsewhere, the sky is considered to be like an umbrella on its pole.[12]

Japan. Out of the primeval oily mass, a reed-like substance emerged. This became a deity, and at the same time two other divine creatures, male and female, came into being. After a period of unmeasured time, a pair of gods called Izanagi and Izanami finally were created. They came down from their heaven to the oily mass by a bridge, generally accepted to have been a rainbow.

Among the offspring of Izanagi and Izanami were geographical landmarks, waterfalls, mountains, trees, herbs, and the wind. They also gave birth to the Sun goddess and the Moon god.[13]

Oceania. From the island of Nauru we hear of a giant clam shell (rather than an egg) in which a light appeared. The upper and lower halves were then separated by a worm (the column), forming the sky and the sea. In Nais, we are told that in the beginning there was a thick fog, which condensed and became a being without speech or movement or head or arms or legs (the cosmic egg). This being in turn gave birth to another, which died, but a tree sprouted from its heart.

From the Society Islands: In the beginning it was said that Ta'aroa developed himself in solitude. He was his own parent and dwelt alone in his shell. The shell was like an egg revolving in endless space, with no sky, no land, no sea, no moon, no sun, no stars. At last he gave his shell a flip and slipped out. But all was silent and dark and he was alone.

He began the task of creation within the darkness. He made trees, plantain clusters, and verdure upon the land. All this was done in a confined world because the sky was still held down upon the earth. Finally the sky was freed and light came into the world.

Gradually the heavens became clear and were decorated with stars. The sun and moon were set in their courses.

Elsewhere: In the beginning there was Po, a void without light, heat, sound, form or movement. From this sort of chaos there gradually developed movement and sound, an increasing light, heat and damp, matter and form, and finally father Heaven and mother Earth.

The tribes of southeastern Australia say that the emu's egg, thrown into the sky, gave birth to the sun by setting fire to a pile of kindling wood. The heavenly deity, recognizing the advantages of this fire, decided to make it burn every day.[14]

South America. The tale of creation as told by the Chibcha Indians of Colombia: "When it was dark, before anything in the world existed, light was enclosed in something big, which the Muiscas called 'Chiminagua.' This thing called Chiminagua -- by which they mean 'god' -- rose to become bright and to free the light that was in it. Then with the emergence of this light, things began to be created."[15]

The similarities of this small sample are obvious. Can a common message be inferred?

The Beginning -- But of What?

Talbott has stated: "The mythical creation is an historical event and has nothing to do with the origins of what we call the natural order."[16] It was instead the beginning of a new order, and various myths detail its unfolding. It was a restructuring of an apparently pre-existing Saturn-Earth system, with untold ramifications for the surface and inhabitants of our planet.

Furthermore, according to Talbott, this reordering was of a celestial, rather than terrestrial, nature. He claims that the translators' rendering of "earth" (as well as other terms) is erroneous, the original having stood for a celestial "abode." This was the organized dwelling of the Great God. Talbott maintains that "the desire to duplicate Saturn's kingdom on earth gave rise to a confusion of terms between terrestrial and celestial dwellings."[17]

The Light of a Thousand Suns

Cardona has observed that a consistent motif in these stories is the Universal, Celestial, Cosmic, or Primordial Egg.

He goes on to state that "we would not be stretching the imagination by positing that the Universal Cosmic Egg was a celestial body . . . [a] spheroid of golden color that mankind saw 'floating' above the waters."

He continues: "A more definitive and direct link connecting the celestial egg with a cosmic body actually exists in an extant Arabic cosmogony of ancient origin which states in no indefinite terms that it was the primordial star that was thus 'presented under the form of an egg that contained all things that were to be born.'"[18]

Cardona then provides evidence that "Saturn, originally a dark sun, went through a fissioning process; that it flared up in a nova-like brilliance; that its emitted light, never to be forgotten by man, blinded the Earth and its inhabitants; that its remains continued to shine as a true sun of night, less bright than that of day but much brighter than the Moon; that mankind witnessed and remembered the event and so had it stated in various texts.''[19]

In addition to the myths that we have cited, we find the phenomenon recorded in Genesis 1:3: "Let there be light." Saturn, "the sun of night," appears in the same chapter, identified as one of the two Great Lights (Genesis 1:16).[20]

These things, if they be true, cannot help but cause wonder, as much with regard to terrestrial conditions prior to the flare-up of Saturn as with respect to the reign and ultimate displacement of the Sun of Night. What does ancient testimony reveal about these matters?

The Age of Twilight

There exist numerous myths and legends that appear to describe the time "before creation."

In New Zealand, we are told, Raugi, the Sky, in love with Papa, the Earth, who was beneath him, came down to her in the time of primeval darkness and immobility. Their close embrace *crushed all beings placed between them*. Nothing could ripen or bear fruit. To escape this situation, the gods determined to separate the sky from the earth. Once the separation was achieved, light spread over the terrestrial world.[21]

The Euahlayi of southeastern Australia say that during a time when there was no sun but only the moon (i.e., a dimmer light) and the stars, a man quarreled with his friend the emu. The emu ran to its nest, took one of its large eggs, and threw it into the sky as hard as he could. There it broke against a pile of wood kindling, which at once caught fire. This greatly astonished *the inhabitants of the earth, accustomed to semidarkness*, and almost blinded them.[22]

From South America we learn that in the most ancient times *the earth was covered in darkness and there was no sun.* For some crime unstated, the people who lived in those times were destroyed by the creator. The sun, moon and stars were created later.[23]

At the head of the Finno-Ugric pantheon stands Jumala, the supreme god, the creator. His sacred tree (the World Tree) was the oak. *His name is related to a word that signifies twilight. or dusk.* It is probable that he was originally a god of the sky, ultimately replaced by another supreme god, Ukko. Ukko was "the ancient father who reigns in the heavens."[24]

The cosmogony of Hawaii speaks of a shadowy void, from which all things emerged, as being the wreck of a preceding world.[25]

Berossus (fl. c. 290 B.C.) apparently speaks of those animals killed during "the creation": "Belus, whom men call Dis, divided the darkness, and separated the Heavens from the Earth, and reduced the universe to order. But the animals so lately created, not being able to bear the prevalence of light, died. Belus upon this, seeing a vast space quite uninhabited, though by nature very fruitful, ordered . . . [the formation of] other men and animals, which should be capable of bearing the light. Belus also formed the stars, and the sun, and the moon. . . .[26]

Pictured in all of the stories is a world whose inhabitants knew no celestial illumination except an eternal twilight; people were totally unfamiliar with the rays of sunlight that bathe our existence today. There was no true daylight and no true darkness. Such an environment could result from the opacity of the atmosphere or from the absence of the sun as we know it, or both.

Earth was a satellite of Saturn, as Velikovsky suggested. Saturn emitted relatively little light (whether always until that time, or just since recently before, we do not know). This low-level light was widely diffused throughout Earth's atmosphere, which was sufficiently clouded, as in fog, to prevent discernment of the great orb providing the dim light from above.

And here we reach *The Beginning*, when Heaven and Earth were one "And the earth was without form and void, and darkness was upon the face of the deep" (Genesis 1:2).

Suddenly, however, the whole countenance of nature began to change.

Revolution of the Natural Order

The body above began to shine with a brighter light. For the first time it became visible as an entity -- the Great Cosmic Egg. And finally it burst forth in a veritable splendor of light, blinding myriads of living eyes accustomed to an existence of semi-darkness.

The first celestial god had been born of himself. He appeared in the heavens as if from nowhere, seeming to emerge from the void. A magnificent column of light would eventually extend toward the earth, to remain and be visible throughout the dawning Golden Age.

According to the Genesis account, day and night were then divided. The light of the great nova was not "daylight"; the night-day cycle, unknown in the previous age, apparently began immediately after the Saturnian flare-up (Genesis 1:4-5). We can infer from this the solar capture of the Saturn-Earth system. It is likely that the repercussions from this capture were in fact responsible for the changes that took place in both Saturn and the earth.

Further atmospheric clearing revealed the *forms* of the Greater and Lesser Lights -- the sun and Saturn (Genesis 1:16).

Saturn was viewed as a beneficent god bearing wonderful gifts for mankind. This is possibly indicative of the world's release from the previously harsher conditions upon the first appearance of the Sun-God.

Thus, man did indeed witness what we have come to call "The Beginning," but his own creation was well before that time.

The two very dissimilar bodies, Saturn and Earth, were destined to orbit the sun as a binary system for untold centuries, or millennia. Far in the future their relationship would be destroyed somehow; the Great Flood would be precipitated from this destruction, and the Golden Age would end. Few people afterwards would realize that the relationship ever existed; uniformitarian assumptions would deny that it ever could have.

The Origin of Planet Earth

Velikovsky claimed that the planet Venus was ejected from Jupiter, subsequently metamorphosing from a comet to its present state.[27] It also has been argued that such is the nature of the origin of comets and other terrestrial-type debris in general;[28] that is, these materials were ejected from Jovian* bodies.

Given the (pre-Deluge) abundance of water (constituents: hydrogen and oxygen) spread over our world, and of hydrogen possessed by Saturn, together with the inferred former association of the two bodies, is it not at least possible that sometime in the very distant, unknowable past, Earth was born of Saturn?#

---

* The giant gaseous "planets" of our solar system: Jupiter, Saturn, Uranus and Neptune.

This contention of course would have to be reconciled with Velikovsky's assertion that the salt content of Earth's oceans came with the floodwaters from Saturn.[29]

REFERENCES

1. Roger W. Wescott, "Aster and Disaster: The Golden Age," *KRONOS* X-l (Fall 1984), pp. 41-51.

2. Immanuel Velikovsky, *Worlds in Collision* (Garden City 1950), passim.

3. Velikovsky, "On Saturn and the Flood," *KRONOS* V-1 (Fall 1979), p. 7.

4. Dwardu Cardona, "On Turning Stones," from the Foreword.

5. David N. Talbott as told to John Gibson, "Saturn's Age," *Research Communications Network Newsletter* #3 (October 15, 1977).

6. Lynn Rose, "Variations on a Theme of Philolaos," *KRONOS* V-1 (Fall 1979), pp. 12-46.

7. Velikovsky, "On Saturn and the Flood," op. cit.

8. Ibid.

9. Veronica Ions, *Egyptian Mythology* (London 1965), pp. 24, 26, 46. Nun later became identified with Hap, god of the Nile. "At a very early period Hap absorbed the attributes of Nun, the primeval watery mass from which the god Ra emerged on the first day of creation. As a result, Hap was regarded as the father of all beings," writes Anthony S. Mercatante in *Who's Who in Egyptian Mythology*, Clarkson N. Potter (New York 1978), p. 51.

10. *New Larousse Encyclopedia of Mythology* (London 1959), pp. 87, 90 .

11. Ions, *Indian Mythology* (London 1967), pp. 25, 28, 29.

12. Anthony Christie, *Chinese Mythology* (London 1968), pp. 49, 53, 57.

13. Juliet Piggott, *Japanese Mythology* (London 1969), pp. 13-14.

14. *New Larousse Encyclopedia of Mythology*, pp. 457, 460, 464.

15. A. Metraux, "South America: Creation and Destruction," *Larousse World Mythology* (London 1972), p. 483.

16. David N. Talbott, op. cit., p. 3.

17. Ibid., p. 4.

18. Dwardu Cardona, "Let There Be Light," *KRONOS* III-3 (Spring 1978), pp. 35, 38.

19. Ibid., pp. 49-50. Also personal correspondence from Dwardu Cardona (May 1989).

20. Cardona, "The Sun of Night," *KRONOS* III-1 (Fall 1977), pp. 31-8.

21. *New Larousse Encyclopedia of Mythology*, p. 462.

22. Ibid., p. 464.

23. Harold Osborne, *South American Mythology* (London 1968), p. 82.

24. *New Larousse Encyclopedia of Mythology*, p. 304.

25. Ibid., p. 460.

26. *Fragments of Berossus*, from Alexander Polyhistor.

27. Velikovsky, *Worlds in Collision*, pp. 173-4.

28. S.K. Vsekhsvyatskii, "The Ring of Comets and Meteorites Encircling Jupiter," *KRONOS* V-4 (Summer 1980), pp. 29-31.

29. Velikovsky, "On Saturn and the Flood," op. cit., p. 8.

CHAPTER XI

The Origin of the Races of Homo Sapiens: A Case of Rapid Divergent Non-Evolution

Race has become almost synonymous with color. We think of people as being red or yellow or black or white. However, since environmental adaptation and interbreeding tend to blur color differences, scientists studying racial variations base their classifications on more concrete identifiers -- less obvious characteristics, perhaps, than color, but traits, nevertheless, that are inheritable and are believed to be unaffected by environment.

According to C. S. Coon, physical anthropologists studying racial differences now rely more on research in blood groups, hemoglobins and other biochemical features than on anthropometry (physical measurements).[1] Blood groups reveal racial differences just as great as the more conspicuous anatomical variations. In fact, biochemistry divides us into the same subspecies that have long been recognized on the basis of other criteria.

Teeth offer another advantage in the study of race, Coon writes. In fact, teeth are just as firmly controlled by genetics as are perishable blood groups. Racial peculiarities in dental details are strictly hereditary.[2]

Walter Karp stresses the importance of such nonadaptive traits as historical tools. When such traits are found in high frequencies in a given race, they are considered to be "race-markers."[3]

Karp also passes along the received opinion on the development of subspecies. To wit: A "race" is a breeding population whose "gene pool" is distinctive from that of other populations. Achieving a distinctive gene pool is no mean feat, however, for unless interbreeding within a population far exceeds outbreeding, a "distinctive genetic heritage" cannot be developed. So how is this heritage brought about? It is asserted that reproductive isolation of different human populations comes chiefly by way of geographical barriers. Such a barrier, Karp states, is the key to the formation of race.[4]

Karp then confesses that the picture of a Negroid race evolving for tens of thousands of years in isolation is belied by the facts. The Sahara Desert, probably the most significant geographical barrier of relevance, is neither ancient nor extremely formidable to gene flow.[5]

Coon maintains that the boundary between Negroes and Bushmen is not impenetrable and that the two subspecies could not have evolved each on its own side of that barrier because the required isolation did not exist.[6]

Where, then, did the dark-skinned people of Africa come from? "The origin of the African Negroes, and of the Pygmies, is the greatest unsolved mystery of racial study," Coon writes.[7]

Caucasians are no less a problem. We are told that they have no clear-cut racial origins.[8]

Meanwhile, far from Europe, nestled in northern Japan, there resides a small and dwindling group of white-skinned people known as the Ainu. These people are a major anthropological enigma. They are thousands of miles from any possible Caucasoid relatives. Their origin is a complete mystery.[9]

Wherever any of these people originated, the accepted time scale of their color differences fits comfortably within the framework of uniformitarianism. We are assured that the "differentiation of white, yellow and black races took place over a period of half a million years by mutation."[10] This assurance is made despite the fact that racial origins remain so much of a mystery.

The pigmentation responsible for skin color serves a definite purpose. In equatorial regions, a higher concentration of pigmentation protects the skin from damage by solar ultraviolet radiation. The synthesis of vitamin D, so useful in the higher latitudes, is enhanced by reduced amounts of pigment. And the conclusion of conventional wisdom regarding this color gradient? It is voiced by C. D. Darlington: "It is therefore evidence of adaptation by natural selection that darker peoples live nearest to the tropics[,] and there is a gradient of skin color correlated with latitude."[11]

But the *ad hoc* nature of this explanation is somehow lost on its adherents. Darlington himself goes on to point out that the correlation is highly erratic; it is lacking in eastern Asia. Even over a range of 40 degrees latitude, adaptation to sunlight is almost absent. With minor exceptions, such as the Ainu, the people there are more homogeneous in all respects than are Western peoples. Darlington says this homogeneity is a result of Eastern peoples' isolation from stocks carrying black pigment and that there is no scope for selective change and adaptation. He extends the same "principle" to the Amerindians and then writes that the contrast between Caucasians and Mongoloids provides a test: "Where color variation is present[,] it is selected and distributed adaptively. But where it is not present[,] it does not easily arise."[12]

This is a remarkable "test." It provides *a priori* results. It looks much like another one of those "plausible stories." Remember the guideline previously cited: If one adaptive explanation fails, try another -- or attribute the failure to some other factor. Surely there must exist more comfortable, less contrived explanations for the origins of racial characteristics.

Dividing Up

There is general agreement that the Amerindians and their Eskimo cousins to the north migrated to the New World from the Old. Compelling evidence for this conclusion is found in the many racial characteristics shared by Amerindians and East Asians.

Coon finds the Amerindians to be fully Mongoloid in skin texture and color range, hair form, hair texture, hair distribution, and degree of sexual dimorphism (differences between the male and female of a given species). He also contends that these two peoples most likely could not have acquired these characteristics independently in Asia and America, that the Asiatic Mongoloids must have acquired them by the time the Amerindian ancestors left Asia.[13]

A consensus holds that the Mongoloid peoples' original passage to the New World was made possible by a bridge of land across what is now the Bering Strait. This ice-free highway supposedly appeared whenever the ocean level was lowered by the formation of ice at the poles during periods of glaciation.[14]

The land bridge of the North was almost certainly the route followed by the "first Americans"; the "Ice Age" and its timing constitute another set of questions. For present purposes, the absolute timing of the crossing is of little or no consequence, but a chronological benchmark may perhaps be relevant to our ideas of how the bridge was inundated. Was the Bering Strait really submerged as the result of an ages-long polar melt? Or is there perhaps another explanation?

The Separation

As noted previously, there exist many records attesting to the historicity of the Great Deluge. Not unexpectedly, the Amerindians have their own recollections of this catastrophe, attributing it to their own gods. Diffusion of Old World memories is not apparent. It thus would appear that the New World was populated prior to the Deluge and that the inundation of the land bridge from Asia resulted from a higher sea level caused by the Flood, rather than from glacial melt.

Given a racial connection between the respective inhabitants of the two hemispheres and their separation as a result of the Flood, certain Mongoloid characteristics must have existed *before* the watery catastrophe. And a question urges itself: Just what did Antediluvian Man look like?

Mirror of the Past

The two most populous racial groupings in the world today are the Mongoloids and other moderately dark-skinned peoples with black hair, some of whom are called Indic. Why should this be? Did all modern races exist somewhere before the Deluge? And were there more survivors in some groups than in others?

More likely, these Mongoloid and other moderately dark-skinned peoples constituted a primal racial group who proliferated after the Flood, before any other groups existed. In effect, they got a head start. And their head start is the likely reason for their making up such a large proportion of mankind today.

Possibly, therefore, these large modern groups bear a strong resemblance to our pre-Deluge ancestors. Perhaps we can even see their reflection in Coon's description of the Mongoloids:

> [T]he Mongoloids of the world, from Madagascar to Tierra del Fuego, are a relatively homogeneous subspecies. They have course, straight, black head hair which grows very long and grays only in extreme senility; and they rarely become bald. Neither sex has very much body hair, and the adult male has very little beard. They have a tendency to facial flatness, protruding malars, widely separated and shallow eye sockets, nasal bones which invade the frontal bone deeply, large incisors which are usually shovelled, relatively long bodies and short lower segments of the arms and legs, along with small hands and feet. . . . Their skin color also tends to vary regionally, but not as much as in the Caucasoid subspecies. Some of them, like the southern Chinese, have very flat noses, whereas others, like the Nagas of Assam and the American Plains Indians, have aquiline ones. But these differences are minimal compared to those found in most other subspecies, and a common origin for all Mongoloids is clearly indicated.[15]

If this yields a rough picture of our remote ancestors, whence the other races?

Let There Be Race

Given the ultimate (pre-Columbian) distribution of the living races, what has been suggested in the way of a primal race forces the conclusion that mankind experienced a basically East (Yellow)-West (White-Black) racial division of its previously anatomically uniform ranks, although certainly with peripheral divagations. The possibility of such a division has been propounded by others.

> In plotting the distribution of [certain] blood-group genes, a striking picture has begun to emerge which has important bearings on the history of race. The deepest differences in blood-group traits lie not between races but between peoples living east and west of the great central Asian mountain desert barrier. On these grounds alone [it has been suggested] that the earliest division of *Homo sapiens* was a differentiation into Eastern and

Western races.[16]

What evidence is there of such a division, and of the further Western latitudinal subdivision into White and Black? And when did this happen?

The mystery of Negro origins, cited above, revolves around their apparent suddenness. When this race of men finally made the scene, they seem to have appeared in a very nonuniformitarian fashion. According to Coon:

> As far as we know, the Congoid line started on the same evolutionary level as the Eurasiatic ones in the Early Middle Pleistocene and then stood still for a half million years, after which [in the Upper Pleistocene] *Negroes and Pygmies appeared as if out of nowhere*.[17]

Additional testimony to the Blacks' sudden appearance is provided by the writings of Ovid, where a recent beginning is implied:

> It was [when Phaethon set the earth aflame], as men think, that the peoples of Aethopia became black skinned, since the blood was drawn to the surface of their bodies by the heat. Then also Libya became a desert, for the heat dried up her moisture.[18] *

The drying-out of the Sahara is believed to have occurred during the Upper Pleistocene or early in the Holocene.

The Ainu also appear to have had a relatively short history, since "[a]ll the known Ainu skulls are recent"[19] (i.e., from the Holocene Epoch) and seem to be an offshoot of their Mongoloid neighbors. "The teeth of modern Ainu are similar to those of prehistorical Japanese, whereas the teeth of modern Japanese are similar to those of ancient Chinese. Despite their Caucasoid appearance, the Ainu are definitely descended from Mongoloid stock."[20]

Sharing with the Blacks of Africa what seems to be a nonexistent

---

* Ovid's allegation of sudden concurrent and compatible changes in man and his environment is thought-provoking. See the discussion relevant to this in Chapter VI.

past, did the European Caucasoids sprout from Mongoloid roots? Karp writes that "European fossils point to the existence of a great diversity of ancient peoples. . . . In Croatia, for example, there are skulls dating back some 100,000 years which have shovel-shaped incisors [a Mongoloid characteristic] and flat 'Mongoloid' faces."[21] And Coon notes: "The teeth of these Europeans of the last interglacial period [Upper Pleistocene] . . . contain morphological features that relate in part to the Sinanthropus-Mongoloid line, unless a common possession of diagnostic Mongoloid features was a coincidence."[22]

A couple of pages back the possibility was raised of an East-West racial division based on blood groups. More specifically, there is a "general resemblance in blood-group traits between Caucasoids and Africans, as well as [an] occurrence in Africa of certain otherwise exclusively Caucasoid blood-group genes."[23] Moreover, as regards racial tooth structure: "The teeth of Caucasoids are plain and simple. . . . Of all races the Caucasoids show the greatest over-all reduction in tooth-pattern detail. . . . Negro and Australian aboriginal teeth are also relatively simple, although less reduced than those of Caucasoids; indeed[,] Negro teeth seem to be most like those of primitive Caucasoids."[24]

A tabulation of results is in order:

1. Black Africans came into existence suddenly in the Upper Pleistocene.
2. The white-skinned Ainu are descended from Mongoloid stock, with no remains known prior to the Holocene.
3. European Caucasoids apparently were preceded by Mongoloid stock in the Upper Pleistocene; fossils from that period have teeth formed on the Sinanthropus-Mongoloid model.
4. There are genetic commonalities between European Caucasoids and Black Africans that are not shared by Mongoloids.

History and the Geological Record

In his ambitious multivolume study of global catastrophes and ancient history, Immanuel Velikovsky has argued for a revised and sometimes compressed chronology of events in antiquity. He elaborated at length on past cataclysms preserved in myths and histories around the world, some of which resulted in a displacement of Earth's axis. Presumably the polar caps were also displaced by partial melting and concurrent extension, evidence of which was later interpreted as indicating "the end(s) of the ice age(s)." Moreover, Velikovsky speculated that some Pleistocene fossils were emplaced -- in historical times -- during one of these disturbances.[25]

At the beginning of the Holocene, the world's ocean levels increased several hundred feet. Supposedly this rise was due to a melting of polar ice at the close of the last ice age. If the presumed "age" did not end, however, in the accepted sense at the accepted time, as suggested above, the Holocene must have commenced before the end of the Pleistocene, which would mean that the additional sea water could not have come from polar ice melt.

It seems likely that the beginning of the Holocene in some parts of the world coincided with, or closely paralleled, the beginning of the Upper Pleistocene in other parts. Both "periods" followed the Middle Pleistocene. (See the proposed Geo-historical Column in the "Afterword.") The first appearance of the Ainu (Holocene) would then fit with that of Caucasoids and Negroids (Upper Pleistocene).

The drastic increase in ocean level must have been brought about by the Deluge. And here is found cause for numerous extinctions assigned to the beginning of the Holocene and for biological changes ascribed to the end of the Middle Pleistocene. Plant mutations at this time might have given rise to such cultigens as maize.

At the beginning of the Middle Pleistocene are also found extinctions and mutations, along with the hologenesis of *Homo sapiens*. I thus assign the Golden Age to the Middle Pleistocene, i.e., to "the Second Interglacial Period," actual duration unknown. This period would have been terminated by the Flood, and soon thereafter mankind was racially divided. Civilization began to emerge in some parts of the world (such as the Middle East), while hunter-gatherers and cave dwellers struggled to survive elsewhere (as in Europe).

The geological evidence of the Flood is far less extensive than is that for earlier upheavals. The Paleozoic, Mesozoic and Tertiary all reflect more violent cataclysms, with devastating wrenchings of the earth. Tidal action, bringing burial and fossilization, was a greater factor in these instances, whereas the Great Deluge, according to legend, was more precipitative in nature. Its effects probably were surfacial (even though Earth's axis may have been tilted so as to initiate or intensify seasons), including such phenomena as the vast Pleistocene soil movements -- as from Canada to the Great Plains of the United

States -- ordinarily attributed to wind and glaciers.

Paleoraces

All of the racial types heretofore mentioned are living today, but there are others less familiar -- less familiar because they are no longer with us. One example is the widely misperceived *Neanderthal Man.*

Ever so anxious for proof of gradualist assumptions, orthodox science has been loath to surrender the evolutionary position of Neanderthal as an ancestral form of man. Neanderthal has proven to be the most divergent form of *Homo sapiens*, a welcome "link" partially bridging the gap from *Homo erectus*.

But the conception of Neanderthal has changed in recent years. Most people still seem to see him as the caveman stereotype, subhuman in nearly every respect. But in his introduction to *The Neanderthals*, R. S. Solecki confesses that scientists now understand Neanderthal to have been not very different from us.[26] They now see the mind of modern man locked into the body of an "archaic" creature. The term "archaic," of course, is very much subjective.

More recently, and more explicitly, we are informed that:

> . . . [T]he status of the Neanderthal -- once viewed as a possible ancestor to modern humans -- has taken a fall. Recent findings have suggested that the Neanderthals coexisted for thousands of years in the Middle East with the forerunners* of modern man. And that means that Neanderthals were not the precursors of *Homo sapiens*.[27]

To this day, the world is populated with many individuals who evidence characteristics typically associated with Neanderthal. Coon writes: "If we compare fossil men such as the Neanderthals with the peripheral, primitive populations of the world, the gap between living and fossil *sapiens* skeletons narrows, until it is closed. Brow ridges reach their peak on Melville Island, and mastoids their minimum in South Africa."[28]

More appropriate, therefore, than "missing link" is the following

---

* It is unclear just what anthropological meaning "forerunners" is supposed to incorporate.

summation with respect to the Neanderthal Race: "[N]othing we know so far rules out the possibility that Neanderthals disappeared when they did not so much through extinction [as] by absorption into the human mainstream. . . ."[29]

Race: No Prospect of an End

There was some degree of human anatomical differentiation through environmental adaptation prior to the Flood, and perhaps even more afterward. But geographical barriers to gene flow were not responsible for the many racial types so familiar today. These types came about not at a uniformitarian pace in the distant past, but catastrophically -- suddenly -- as the result of widespread, and probably simultaneous, transmutations (cause unknown) -- and this within man's memory. From a monotonous homogeneity, there quickly emerged a veritable rainbow of humanity: fair skin and light-colored eyes and golden hair to black skin and black hair that was no longer straight, i.e., distinct racial types.

Moreover, race came first. The barrier maintaining it followed. That barrier is universal: It is socio-psychological.

REFERENCES

1. C. S. Coon, *The Origin of Races*, Alfred A. Knopf, Inc. (New York 1962), p. 622.
2. Ibid., p. 351.
3. W. Karp, "How Did Human Races Originate?" *Mysteries of the Past*, ed. by J. J. Thorndike, Jr., American Heritage (New York 1977), p. 217.
4. Ibid., p. 213.
5. Ibid., p. 225.
6. Coon, op. cit., p. 589.
7. Ibid., p. 588.
8. *Encyclopaedia Britannica*, 15th ed., Vol. 6, p. 1051.
9. "Who Are the Hairy Ainu?" *Mysteries of the Past*, op. cit., p. 231.
10. C. D. Darlington, *The Evolution of Man and Society*, Simon and Schuster (New York 1969), p. 44.
11. Ibid., p. 43.
12. Ibid.
13. Coon, op. cit., p. 480.
14. Darlington, op. cit., p. 577.
15. Coon, op. cit., p. 428.
16. Karp, op. cit., p. 219.
17. Coon, op. cit., p. 658 (emphasis added).
18. Ovid (43 B.C.-c.A.D. 17), *Metamorphoses* II:234-7.
19. Coon, op. cit., p. 477.
20. "Who Are the Hairy Ainu?" op. cit., p. 231.
21. Karp, op. cit., p. 228.
22. Coon, op. cit., p. 27.
23. Karp, op. cit., p. 225.
24. Coon, op. cit., p. 363.

25. Immanuel Velikovsky, "Ash -- A Historical Record," *Pensée* IV-1 (Winter 1973-1974), p. 6.

26. R. S. Solecki in G. Constable, *The Neanderthals*, Time-Life (New York 1973), p. 6.

27. "Pity the Poor Neanderthal," *Science Impact* 1-12 (May 1988), p. 3.

28. Coon, op. cit., p. 346.

29. "Pity the Poor Neanderthal," op. cit.

CHAPTER XII

The Tower of Babel and the Catastrophic Non-Origin of Language Diversity

And the whole earth was of one language and of one speech.

And it came to pass, as they journeyed from the east, that they found a plain in the land of Shinar; and they dwelt there.

And they said to one another, Go to, let us make brick, and burn them thoroughly. And they had brick for stone, and slime had they for mortar.

And they said, Go to, let us build us a city and a tower, whose top may reach unto heaven; and let us make us a name, lest we be scattered abroad upon the face of the whole earth.

And the Lord came down to see the city and the tower, which the children of men builded.

And the Lord said, Behold, the people is one, and they have all one language: and this they began to do: and now nothing will be restrained from them, which they have imagined to do.

Go to, let us go down, and there confound their language, that they may not understand one another's speech.

So the Lord scattered them abroad from thence upon the face of all the earth: and they left off to build the city.

> Therefore is the name of it called Babel; because the Lord did there confound the language of all the earth: and from thence did the Lord scatter them abroad upon the face of all the earth.
>
> Genesis 11:1-9

Confusion Upon Confusion

Unlike the worldwide prolificacy of flood traditions, extant legends of the *Confusion of Tongues*, such as that recorded in the biblical story of the *Tower of Babel*, are much fewer in number. The skeptic might still discern a kernel of truth in the flood legends even while rejecting the idea of a global deluge, but stories of the Confusion of Tongues are more easily relegated to the realm of fairy tales. Even so, their widespread and spotty distribution begs a historical explanation that can preclude their having resulted from a single source. An analysis of some of these stories also suggests not only that they are not fairy tales, but that our usual conception of what they describe suffers from more confusion than did mankind's speech.

In an earlier chapter, I speculated on the early development of different languages. This chapter answers the anticipated creationist response to that assertion.

The biblical account of the Tower of Babel actually incorporates two themes found in mythologies scattered around the world: the Confusion itself and man's attempt to scale heaven. In the extrabiblical versions, these two themes occur sometimes independently and sometimes together (see the appended matrix). Our primary interest is the Confusion, but certain aspects of those scaling myths not explicitly associated with the theme of Confusion are nevertheless of interest as well.

The widespread occurrence of the scaling theme would seem at first to be a product of diffusion, reflecting an ancient human endeavor of sorts rather than a global disturbance. However, one logical connection will be suggested to connect the catastrophe with a structure used for scaling the heavens.

Interpretation of the Records

The familiar biblical account of the Confusion of Tongues unquestionably was recorded at an early date. Most of the other Confusion legends, until recently, were transmitted orally, with the reliability of their details inevitably suffering over the centuries. With this in mind, there is to be noted a possible misunderstanding of the nature of the Confusion as stated in Genesis.

From earlier than the time of Josephus (c. A.D. 37-c. 100), this record has been interpreted as declaring an instant diversification of human language, i.e., the miraculous creation of new languages. In the words of Josephus: "[God] created discord among them by making them speak different languages, through the variety of which they could not understand one another."[1]

This belief still prevails. Creationists point to the Confusion as one of the greatest miracles recorded in the Old Testament. Herbert Lockyer writes: "The true origin of the diversity of languages and distinct dialects, of which there are now over 7,000, is here in the miracle before us."[2]

Skeptics agree with this interpretation, but reject its historicity. According to Isaac Asimov, "God defeated [the builders'] purpose by giving each man a different language, making it impossible for them to understand each other."[3] An authority writes in the *Encyclopaedia Britannica*, "The mythical story of its [Tower of Babel's] construction . . . appears to be an attempt to explain the existence of diverse human languages."[4]

But a careful examination of the biblical text reveals no explicit statement that such a phenomenon occurred. The passage in the eleventh chapter of Genesis reads in part:

> And the whole earth was of one language, and of one speech.
> . . . Go to, let us go down, and there confound their language,
> that they may not understand one another's speech.
> . . . Therefore is the name of it called Babel; because the
> Lord did there confound the language of all the earth.*

The idea of language diversification is not present in the account. The text simply states that the builders' speech was *confounded*; it is not suggesting that groups of people became unable to communicate with each other because of an instantaneous language change. Rather,

---

* Babel -- "Gate of God": It was called Babel not because God *confounded* the language there, but because God confounded the language *there*.

no one affected could communicate with anyone else; the power of coherent speech was temporarily lost.

The condition proclaimed in Genesis 11:1 can be read as the prevailing situation in Southwest Asia at that time; that is, there was *one* language before, during and after the disturbance. The purpose of this verse probably was to emphasize the resulting consternation over the fact that, even though the victims spoke the same language, they could not communicate orally.

Josephus' interpretation is possibly contradicted through his own reference to another source:

> This tower and the confusion of the tongues of men are mentioned also by the Sibyl in the following terms: "When all men spoke a common language, certain of them built an exceeding high tower, thinking thereby to mount to heaven. But the gods sent wind against it and overturned the tower and gave to *every man* a peculiar language; whence it comes that the city was called Babylon."[5]

One wonders how literally to take the Sibyl. If "every man" can be equated with "every single individual," an import that seems implied, this reading is additional support for interpreting "peculiar language" or "confusion of tongues" quite simply as total confusion.

Total confusion is more than intimated in Jewish legend; there we find it stated that "none [that is, not a single person] knew what the other spoke. One would ask for the mortar, and the other handed him a brick. . ."[6]

A turn-of-the-century translation of the Assyro-Babylonian account, the oldest known extrabiblical version, makes no allusion to new languages:

> Babylon corruptly to sin went and small and great mingled on the mound.
> The King of the holy mound. . . .
> In front and Anu lifted up . . . to the good god his father.
> . . . Then his heart also . . . which carried a command.
> . . . At that time also . . . he lifted it up. . . .
> Their (work) all day they founded to their stronghold in

the night entirely an end he made.
In his anger also the secret counsel he poured out to scatter (abroad) his face he set *he gave a command to make strange their speech.*
. . . their progress he impeded. . . .
In (that day) he blew and . .
For future time the mountain.
Nu-nam-nir went . . .
Like heaven and earth he spake.
His ways they went. . . .
Violently they fronted against him.
He saw them and to the earth (descended)
When a stop he did not make of the gods . .
Against the gods they revolted . . . violence.
Violently they wept for Babylon very much they wept . .[7]

The content of the other extrabiblical legends varies. Some tell of the creation of new languages; others, however, tell only of a confusion or confounding.

One might expect that the popular interpretation of language diversification would arise eventually, explicitly in some of the legends and by extrapolation into others. A foreign language, after all, sounds like gibberish to an untrained ear. Might not the same listener assume gibberish to be another language? The biblical version, however, manifestly was recorded prior to the interjection of this thought.

Contemporary Parallels

The new interpretation offered above is supported in a number of records of modern scientific observation which suggest that the Confusion quite literally may have been a shocking experience:

> Application of an electric current across a group of neurons elicits both excitatory and inhibitory effects. *Vocalization can be evoked by electrical stimulation of the motor strip, but these vocalizations are never words.* Spontaneous language has not been evoked from cortical stimulation. Rather, cortical stimulation seems to act on such complex behavior as language as though it were introducing noise into the system. The

> brain apparently does not have time to compensate for this sudden burst of noise, so its introduction at any point in the complex system involved in language production may interfere with ultimate performance. . . . *The period of disruption of function is, in large measure, temporary.*[8]

Another authority writes:

> The "arrest response" produced by [electrical] stimulation in the [brain's] posterior frontal region . . . consists in the arrest of voluntary movement and may show additional features such as *post-stimulation confusion, inappropriate or garbled speech*, overt mood changes. . . .[9]

The elicitation of such effects is not restricted to direct electrical stimulation:

> ELF [Extremely Low Frequency] electromagnetic fields and waves *may be important biological stimuli* because of their penetrability and long distance propagation. . . . [T]heir frequencies and intensities are within the ranges of processes generated by living organisms.[10]

ELF fields may affect people in several ways: "[D]iffuse behaviors, such as ambulation or emotional responses, . . . have been reported to vary as a function of ELF electric or magnetic field application."[11]

Thus, small-scale repetitions of the ancient event are produced from physiological disturbances -- by shock and possibly by exposure to electromagnetic radiation. As with the modern experiences, the ancient catastrophic effects (aside from possible death) were transient, and afterwards the victims again could communicate in their own language.

We may now envisage a progression of events: The capability of coherent speech was temporarily interrupted; the effects of the catastrophe, coupled with the confusion and fear that resulted, forced a dispersion of the people. Interesting to note is that, even in the modern instance, "electrical stimulation with subconvulsive intensities . . . elicits . . . *emotions of rage and fear*."[12]

Language diversification, an ongoing process around the world from the very inception of *Homo sapiens*, then would have proceeded at its own *uniformitarian* pace. This difference in language would have been recognized much later. Within memory of the confusion, however, and with little else comprehensible to men of that time to account for language differences, the Confusion at Babel understandably would have been held responsible by many.

As chance would have things, there exists in African legend a somewhat surprising similar theme to the foregoing conjecture: "The Wa-Sania of British East Africa say that of old all the tribes of the earth knew only one language, but that during a severe famine *the people went mad and wandered in all directions, jabbering strange words*, and so the different languages arose."[13]

Hebrew legend suggests another unusual aspect of the phenomenon, reflecting the incomprehensible disorientation associated with the Tower's site: "The place of the tower has never lost its peculiar quality. Whoever passes it forgets all he knows."[14] And again, from our own time: "A . . . phenomenon during thalamic [electrical] stimulation is monosyllabic yells and exclamations . . . which seem to be utterances of surprise, fright, or pain. . . . *Most of the patients who produced [such] speech were unable to recollect what was said or that anything had been said, though they are conscious immediately after the end of the stimulation.*"[15]

The Toppled Tower

There is a curious aspect to the legends, the significance of which is uncertain. All of the scaling legends that do not involve the Confusion of Tongues cite the collapse or destruction of the structure involved. In some legends, this resulted in the death of its builders. In the tales that include both the Tower and the Confusion, such as the biblical account, the fate of the Tower is not always revealed. Perhaps this omission merely reflects a shift of narrative interest in the stories containing both themes. Nevertheless, the destruction seems to be an integral part of mankind's memory of those time-shrouded events.

The Tower Toppler

If the Confusion resulted from shock and/or exposure to electromagnetic radiation, there must necessarily have been some natural means by which it might have occurred. According to M. A. Persinger, there is indeed a potentially destructive agent with the capacity to wreak the Confusion of Genesis 11. "[O]ne can infer that ELF and VLF signals with wave characteristics have their origin almost exclusively in *lightning strokes*," Persinger writes, "even though relationships to disturbances in the earth's magnetic field as well as the ionosphere and more outward layers are known."[16]

In legends are found at least two references to the destruction of the Tower by lightning. From the Old World:

> Giants attacked the very throne of Heaven,
> Piled Pelion on Ossa, mountain on mountain
> Up to the very stars. Jove struck them down
> With thunderbolts, and the bulk of those huge bodies
> Lay on the earth. . . .[17]

And from the New World:

> In Mexico, it is told of the great pyramid of Cholula that once upon a time certain giants aspired to build out of clay and bitumen a tower which would reach to heaven, so that they might enjoy from the top of it the spectacle of the rising and setting sun. When, however, they had reared it as high as they could, the inhabitants of heaven, at the bidding of their overlord, came down to all four quarters of the earth, overthrew the structure by a thunderbolt, and scattered the builders in all directions.[18]

There are two other possible allusions to lightning. From Africa:

> The Njamwezi of Tangyika [say that the] Valengo, a large family which lived in primeval times, decided one day to build a tower to heaven. For several months they worked assiduously on the project, until at last it seemed to be nearing completion. Thereupon they summoned all their sons and grandsons to witness the crowning moment. Suddenly, however, a storm burst from the sky and overturned the structure. The Valengo were killed to the last man. . . .[19]

And from the Hittites:

. . [T]he god Kumarbi, at odds with the reigning sovereign of heaven, engendered a monstrous creature, Ullikummi, who grew up miraculously *in the form of a stone pillar* and, mounted on the shoulder of a subaqueous giant, tried to rock the floor of heaven and topple the gods from their thrones. In this case too, however, the impious effort proved abortive, for the god Ea severed the monster from his support by means of a magic cleaver [electrical discharge?], and he crashed forthwith to his doom.[20]

Although there is no reference to lightning in the biblical account, Jewish legend does have it that the Tower of Babel experienced some kind of pyrogenic assault. "As for the unfinished tower," writes Louis Ginzberg, "a part sank into the earth, and another part was consumed by fire; only one-third of it remained standing."[21]

Conclusions About Conclusions

An objective weighing of the biblical record, legendary material, and contemporary experience together suggest that the catastrophe that befell the Tower of Babel probably took the form of a thunderbolt. The only strange language involved seems to have been "garble-ese."

Since diffusion could not likely have accounted for all of the legends, the phenomenon could not have been restricted to Mesopotamia, but must have occurred elsewhere as well.

The Tower perhaps acted as a lightning rod.* And therein is a possible reason for the legendary association of such towers with the Confusion. Societies with an inclination to build tall towers at that time in the earth's history might have stood a higher probability of being "zapped." (One can be pardoned for wondering why societies might have had "an inclination to build tall towers." Perhaps they were trying to imitate some image that they saw in the heavens. Or perhaps they were trying to reach such an image. There are numerous possible solutions, none of them provable.)

Also of sufficient importance to note well is the fact that some of

---

* This was the opinion of Immanuel Velikovsky. He also suggested in an unpublished manuscript that the discharge was interplanetary in nature.

the myths do not quite fit the interpretation given above. In fact, Dwardu Cardona has a very different explanation: "Ancient texts . . . more than intimate that the Tower was not a man-made edifice but, rather, a celestial apparition that was nonetheless physical."[22]

In pointing to this event as the origin of the world's many different languages, the creationists have again misinterpreted their source of information. And now that the account need no longer be viewed as incredible, can its historicity not be accepted? Or is it, by habit, to continue to be rejected out-of-hand as just another "biblical fairy tale" -- or as a violation of an unyielding philosophy?

FREQUENCY DISTRIBUTION OF ELEMENTS WITHIN THE ACCOUNTS ON WHICH THIS STUDY IS BASED

	Artificial (Tower)	Natural Structure	No Reference	Total
Confusion *	4	1	1	6
New Languages	4	2	8	14
No Reference	12 †	0 †	-	12
TOTAL	20	3	9	32

* No specific reference to new languages.

† An extensive search of myths for this element has not been made.

BIBLIOGRAPHY FOR THE ABOVE MATRIX

Genesis 11:1-9

Flavius Josephus, *Jewish Antiquities* I:109-121, trans. by H. St. J. Thackeray.

W. St. Chad Boscawen (trans.), *The World's Greatest Literature* (N.Y 1901), pp. 233-4.

T.H. Gaster (ed.), *Myth, Legend and Custom in the Old Testament* (N.Y. 1969), pp. 132-8.

L. Ginzberg. *Legends of the Jews* (N.Y. 1956), Vol. I, pp. 84-5.

Ovid. *Metamorphoses* I:153-7.

E.E. Clark, *Indian Legends of the Pacific Northwest* (Los Angeles 1960), pp. 43, 138-141.

Larousse Encyclopedia of Mythology, (London 1959), pp. 438, 440.

B. Nelson, *The Deluge Story in Stone* (Minneapolis 1931), pp. 183-7.

REFERENCES

1. Flavius Josephus (c. A.D. 37-c. 100), Jewish Antiquities, I:117.
2. Herbert Lockyer, *All the Miracles of the Bible*, Zondervan (Grand Rapids 1961), p. 35.
3. Isaac Asimov, *Asimov's Guide to the Bible*, Doubleday (Garden City 1971), p. 54.
4. *Encyclopaedia Britannica*, 15th ed., "Tower of Babel," Index I, p. 707.
5. Flavius Josephus, *Jewish Antiquities*, I:118 (emphasis added).
6. Louis Ginzberg, *Legends of the Jews*, Vol. 1 (New York 1956), pp. 84-5 (emphasis added).
7. W. St. Chad Boscawen (trans.), *The World's Greatest Literature* (New York 1901), pp. 233-4 (emphasis added).
8. H. and H.A. Whitaker, *Studies in Neurolinuistics*, Vol. 1 (New York 1976), pp. 120-1 (emphasis added).
9. J. Van Buren, C. L. Li, G. Ojemann, "The Fronto-Striatal Arrest Response in Man," *Electroencephalography and Clinical Neurophysiology* 21 (1966), p. 128 (emphasis added).
10. M. A. Persinger, *ELF and VLF Electromagnetic Field Effects* (New York 1974), p. 4.
11. Ibid., p. 276.
12. E. Gellhorn and G. N. Loofbourrow, *Emotions and Emotional Disorders* (New York 1963), p. 314 (emphasis added).
13. T. H. Gaster, ed., *Myth, Legend and Custom in the Old Testament*, Harper and Row (New York 1969), p. 136 (emphasis added).
14. Ginzberg, op. cit., p. 85.
15. G. Schaltenbrand, "The Effects on Speech and Language of Stereotactical Stimulus in Thalamus and Corpus Callosum," *Brain and Language* 2 (1975), pp. 70-7 (emphasis added).
16. Persinger, op. cit., p. 11 (emphasis added).
17. Ovid, *Metamorphoses* I:153-7.
18. Gaster, op. cit.

19. Ibid., p. 133.

20. Ibid., p. 135 (emphasis added).

21. Ginzberg, op. cit., p. 85.

22. Dwardu Cardona, "The Road to Saturn," Part II, *Aeon* I-3 (March 1988), p. 119.

CHAPTER XIII

Sodom, Gomorrah and the Electrological Origin of the Dead Sea Depression

Sodom and Gomorrah are probably the two most infamous cities that ever existed, the name of one having established itself in the English language in a term evidencing an undying image of wickedness: *sodomy*. Although their notorious image persists, the cities themselves have been lost for millennia, and only recently have their sites been tentatively identified.[1]

The Dead Sea is contained within a sunken block confined by two parallel geological faults, a configuration indicative of a catastrophic origin.[2] The depression was not created by the action of water, and the "sea" is not really a sea but is a landlocked lake. We are told that during the Miocene Epoch (from 7 million to 26 million years ago), upheaval of the Mediterranean Sea bed produced the upfolded structures that caused the fractures forming the Dead Sea depression.[3]

The destruction of Sodom and Gomorrah was probably not an isolated event. Their fate, attested by more than one ancient record, provides one more bit of testimony for the further verification of the belief that Earth has not been the insular, stable planet usually typified and that it has been scarred from above by more than just falling rocks.

The Book of Genesis tells of Abraham's kinsman Lot electing to dwell near Sodom (Gen. 13:12) and of his escape prior to the destruction of the two cities (Gen. 19:16), his wife at that time being transformed into a *pillar of salt* (Gen. 19:26). The destruction is said to have resulted from fire and brimstone falling from heaven (Gen. 19:24). The fact of the catastrophic destruction of Sodom and Gomorrah is sometimes acknowledged by scientists and historians,[4] although the means is considered to have been of a purely geological nature.[5] The biblical description of the destruction is regarded typically as either a misconception or a fabrication.

One Arabic name for the Dead Sea is *Sea of Lot* (*Buhayrat Lut*),[6] possibly a remembrance that the depression was formed not eons ago but as a consequence of "fire from heaven" in the time of Lot.* An apparent parallel in Scandinavian mythology comes to mind: terrestrial valleys said to have been dug by Thor's hammer. An incident in the life of Abraham as preserved in Hebrew legend perhaps suggests as much: "The day whereon God visited him [Abraham] was exceedingly hot, for he had bored a hole in hell, so that its heat might reach as far as the earth. . ."[8]

One must presume that the destruction of the cities would have been attributed by biblical scribes to a terrestrial upheaval, such as volcanic action or an earthquake, with no recourse to a celestial agent as an explanation for the event, unless there was compelling eyewitness testimony to that effect. What, however, might have been the agent, and what was the nature of the fire?

Velikovsky has suggested that Jupiter was once a much closer neighbor than it is today. And this planetary behemoth was not a passive body, but was active to the point of being the progenitor of a number of terrestrial troubles. Once this thought is entertained, Velikovsky's contention should become an obvious interpretation to any reader of ancient mythology.

Accounts of the exploits of the God of Thunder and Lightning are found in cultures worldwide. Appellations for the god vary, but their contexts offer a common testimony. The less fanciful accounts make clear the fact that the planet Jupiter was the Thunder God. He is frequently described as hurling his bolts to the earth:

---

* But the present extent and depth of the lake probably resulted from a later catastrophe during the time of Moses.[7]

> Lo, through the clouds the father of the gods scatters red lightnings, then clears the sky after the torrent rain: never before or since did hurtling fires fall thicker . . . terror filled the hearts of common folk. . . . It is possible to expiate the thunderbolt, and the wrath of angry Jove can be averted.
>
> . . . [T]he sun had already lifted his full orb above the horizon, and a loud crash rang out from heaven's vault.
>
> Thrice did the god thunder from a cloudless sky, thrice did he hurl his bolts.[9]

Myth, legend and ancient history all attest that this same Jupiter once dominated our heavens as a brilliant celestial light. Hebrew legend is very specific in regard to Jupiter's former proximity and brilliance, and it also suggests a chronological context. "Abraham's [military] victory was possible only because the celestial powers espoused his side," Louis Ginzberg writes. "The planet Jupiter made the night bright for him[,] and an angel, Lailah [Hebrew: *night*] by name, fought for him.''[10]

Velikovsky asserted that in the era of Jupiter's prominence in our skies, Earth witnessed the Thunder God's propagation of Venus -- the ejection of primal planetary matter by a giant of the solar system -- Venus being thrust into an orbit intersecting that of Earth. Supposedly, the Greeks came to see this event as the birth of Athene from the head of Zeus.[11]

Whatever the past interrelationships of Jupiter and Venus, these planetary deities constituted a menacing presence in the vault above, ultimately making the Cities of the Plain victims of an unearthly disaster.

According to Velikovsky, "The planet-god Jupiter . . . was pictured with a thunderbolt because of the spectacles witnessed by the inhabitants of the earth -- like a discharge that was directed toward Venus when it approached its parental body."[12] Moreover, "the phenomenon of brimstone (sulphur) falling from the sky . . . in the course of great discharges, as narrated in ancient sources (Old Testament and Homer among them), resulted from smashing two oxygen atoms into one atom of sulphur,"[13] sulphur being a notable constituent of the atmospheres of both Jupiter and Venus.

Sodom and Gomorrah apparently were target zero for one such discharge. The lovely plain was burned, gouged and jolted -- never to be the same again -- as the cities and their inhabitants were incinerated in an electrical torrent and carried into their common grave. Lot's wife was possibly a victim of the radiation emitted by the stupendous arc, "crystallized" (for lack of a better term at the present) as she watched the annihilation unshielded. The destruction was so widespread and so complete that Lot's daughters thought themselves and their father to be the last people left alive upon the earth (Genesis 19:31-38).

Jupiter's ragings were feared elsewhere in the Middle East as well: ". . . Enlil, [Sumerian] god of the air . . . stood for force . . . the thunder and lightning over the Mesopotamian plain was Enlil, and also . . . it was Enlil . . . who destroyed a city."[14] The author reporting this incident confessed that he was unable to fathom the "true meaning" of the record.

Although the era is uncertain, we find a striking parallel to the biblical record in an account from South America: "[A]t Pucara, which is forty leagues from Cuzco on the Collao road, fire came down from heaven and burnt a great part of them while those who tried to escape were turned into stone."[15]

Earth history and topography thus bear more than one type of celestial imprint. What additional strange phenomena and origins remain hidden from our knowledge because of our preconceptions and philosophies?

REFERENCES

1. See discussion by Dwardu Cardona, "Jupiter - God of Abraham" (Part III), *KRONOS* VII-1 (Fall 1982), pp. 69-71.
2. *Encyclopaedia Britannica*, Vol.5, 15th ed., p. 524.
3. Ibid.
4. Ibid., Index IX, p. 322. See also Werner Keller, *The Bible as History*, Morrow (New York 1956), p. 76.
5. *Britannica*, p. 322. Keller, p. 76.
6. *Britannica*, Vol. 5, p. 524.
7. See discussion by Immanuel Velikovsky, "The Destruction of Sodom and Gomorrah," *KRONOS* VI-4 (Summer 1981), p. 44.
8. Louis Ginzberg, *Legends of the Jews*, Jewish Publication Society (Philadelphia 1913), p. 240.
9. Ovid, *Fasti* III:285-91, 369-70.
10. Ginzberg, op. cit., p. 232.
11. Immanuel Velikovsky, *Worlds in Collision*, Doubleday (Garden City 1950), pp. 156-9.
12. Velikovsky, "On the Advance Claims of Jupiter's Radio Noises," *KRONOS* III-1 (Fall 1977), p. 27.
13. Velikovsky, "Venus' Atmosphere," *Pensée* IV-1 (Winter '73-'74), p. 35.
14. W. G. Sinnigen and C. A. Robinson, Jr., *Ancient History* 3rd ed., Macmillan (New York 1981), p. 28.
15. Harold Osborn, *South American Mythology*, Hamlyn (London 1968), p. 64.

CHAPTER XIV

Archaeo-Electrics: A Once-Occasional Factor In the Origin of Holy Writ

Glow discharges of various kinds have long been of interest to man, no doubt evoking a great deal of wonder in times past. The forms of discharge are so varied that it has been only during the present century that they have been shown to be different aspects of the same basic physical phenomenon: In every form, the glow discharge is a manifestation of the conduction of electricity through gases. One form occurring in nature is the faint glow known as St. Elmo's fire, seen at the top of ships' masts at night when thunderclouds are near. Other natural forms are seen from sharp Alpine peaks and even from the heads of human beings. The edges of aircraft wings sometimes glow when passing through thunderclouds. Another manifestation is the *Aurora Borealis*, the so-called Northern Lights.[1]

Because the rounded head of the ship mast wore the light of St. Elmo's fire like a crown or halo, this phenomenon was later called *corona*, the Latin word for crown. As sources for high-voltage electricity were developed many years later, the same light-like phenomena were observed in the laboratory and also called corona. Such phenomena produce not only light, but audible sound as well.[2]

Corona, or partial discharge, is defined as a type of discharge resulting from the ionization of atoms in an electric field, a condition caused by a variety of physical conditions within the earth and/or its atmosphere. If the associated voltage becomes great enough, arcing will occur; lightning is a familiar example. As suggested above, corona discharges do not necessarily require electrically conducting materials.[3,4]

In our generation, we all have grown up in the midst of electrical technology. We have no difficulty in distinguishing luminous electrical phenomena from the chemical luminosity of fire. But suppose that electricity were unknown except through its natural manifestations. How would ancient man have described a corona discharge? Most certainly as fire -- the process of combustion. By accepting his description of an observation merely at face value (accurate words for the phenomena in question did not exist), we might be led to believe that a statement of simple fact is totally incredible. To arrive at the truth, we must adjust our frame of reference and see as ancient man saw, think as he thought. Distinguishing between electricity and combustion makes credible a number of records he left us.

It is the credibility of the ancients that underpins this chapter. If the ancients cannot be believed, we are thrust entirely upon our own devices in piecing together our antecedents. If they *can* be believed, our position is far better; we have ancient partners in our attempts to understand ourselves.

One thing we can understand today that the ancients could not comprehend is the extent to which our world (as well as our own bodies and minds) is electrical. Man's belated knowledge of electricity ultimately may permit a reconstruction of a time in human history when electrical phenomena were truly spectacular.

Atmospheric electric fields are commonplace today; however, various ancient descriptions of strange fire suggest that in the past these fields achieved magnitudes far greater than is usual today. Velikovsky postulated an electrical discharge from the ionosphere to the earth as a result of a localized increase in ionospheric charge.[5] A tremendous electric field, resulting from the charge, would have preceded such a discharge. Destructive manifestations did not always result, but extraordinary fields nevertheless could announce their presence through the less violent corona.

These and related natural electrical phenomena seem to explain a number of otherwise mysterious events recorded in the Old Testament.

We look back some 3,000 years, to the exodus of the Israelites from their bondage in Egypt.

The Tabernacle of Israel

Soon after the Sinai spectacle -- *The Lawgiving* -- still at the very beginning of their forty-year wandering, the newly liberated Israelites undertook the construction of a center of worship: the Tabernacle. It was a sizable structure but nevertheless portable, designed to be readily assembled or disassembled. The aspects of the design are given in excruciating detail in the Book of Exodus.

Much of the Tabernacle was made of fabric or hides, but there was also an astounding amount of metal:

29 talents and 730 shekels of gold (Exod. 38:24);
100 talents and 1,775 shekels of silver (Exod. 38:25);
70 talents and 2,400 shekels of copper (Exod. 38:29).

In addition to various riggings and utensils of gold, silver and copper, there were over fifty panel frames some 10 x 1.5 cubits in size overlaid with gold (Exod. 26:29). The Ark of the Covenant (Exod. 25:11) and the Table for holding the Showbread (Exod. 25:24) also were overlaid with gold, and the Altar was overlaid with copper (Exod. 27:2). This assemblage occupied an area of some 5,000 square cubits (Exod. 27:18) -- its dimensions and bulk were totally foreign to its nomadic context.

Yet this huge Tabernacle dictated the pace of the children of Israel in their wanderings, not merely in its bulkiness, but in seemingly supernatural phenomena that told Moses and his followers whether to move or stay put. From the time of its completion, the Tabernacle was a focal point of remarkable phenomena: "So Moses finished the work. And the cloud began to cover the tent of meeting, and Jehovah's glory filled the tabernacle. And Moses was not able to go into the tent of meeting, because the cloud resided over it and Jehovah's glory filled the tabernacle" (Exod. 40:33-35).

According to Exodus 29:42-45, the Lord told Moses that He would "dwell among the children of Israel" in the Tabernacle. What came to be known as "the Shekinah glory" was the Israelites' sign that God was in the Tabernacle, and the presence or absence of this cloud determined whether the people of Israel would encamp to worship the Lord or pull up stakes and continue their journey: "And when the cloud lifted itself up from over the tabernacle, the sons of Israel would break camp during all their stages of journey. However, if the cloud did not lift itself up, then they would not break camp until the day when it lifted itself up" (Exod. 40:36-37).

The appearance of the Cloud depended on the time of day: "For Jehovah's cloud was over the tabernacle by day, and a fire continued upon it by night in the sight of all the house of Israel during all their stages of journey" (Exod. 40:38).

This was not the pillar of fire/cloud said to have moved ahead of the Israelites (Exod. 13:21-22) and identified by Velikovsky as a comet.[6] This was no comet in the sky. It was earthbound and intermittent and associated strictly with the Tabernacle. It was, to some degree, approachable, but coming into its presence could have strange consequences.

> And he [Moses] went gathering seventy men from the older men of the people and proceeded to have them stand round about the tent. Then Jehovah came down in a cloud and spoke to him and took away some of the spirit that was in him and put it upon each of the seventy older men. And it came about that as soon as the spirit settled down upon them, then they proceeded to act as prophets; but they did not do it again (Num. 11:24-25).

The Seventy apparently either appeared to go into a trance (as a mystic might) or else became incoherent in some aspect of their behavior -- a "prophetic rapture." This same effect was experienced by two others in the camp further away from the tent (Num. 11:26).

There has been a considerable amount of speculation about the possible electrical properties of the Ark of the Covenant. Because of its construction, the Ark (as well as other items in the Tabernacle) would appear to have had the ability to store an electrical charge.* A conductive metal overlay on both sides of a nonconductor is descriptive of a device commonly employed for such a purpose in modern

---

* Recorded in II Samuel 6:7 is a death attributed to contact with the Ark -- perhaps an electrocution as the Ark was grounded through a human body.

electrical technology: a capacitor.

The Ark also was described as a medium of communication: "Now whenever Moses went into the tent of meeting to speak with him [Jehovah], then he would hear the voice conversing with him from above the cover that was upon the ark of the testimony, from between the two [gold] cherubs; and he would speak to him" (Num. 7:89).

The Voice of the Ark did not constitute Moses' first contact from Jehovah. Earlier, Moses had been commissioned in a similar fashion:

> Then Jehovah's angel appeared to him in a flame of fire in the midst of a thornbush. As he kept looking, why, here the thornbush was burning with the fire and yet the thornbush was not consumed. . . . God at once called to him out of the midst of the thornbush and said: "Moses! Moses!" (Exod. 3:2,4).

What is to be made of all this?

As to the Burning Bush: A bush that burned without being consumed was not on fire. Broached in the second and third paragraphs of this chapter was a phenomenon called *corona*, a discharge resulting from the ionization of atoms in an electric field. The Burning Bush scene was repeated in 1966 when there was "experimentally demonstrated a . . . corona discharge from a small bush 46 cm (18 in) in height which was placed in [an electric] field."[7*]

"Burning-bush red" is not the extent of the corona repertoire, as "the color in discharge manifestations in air [varies; it] is a guide to [electric] field conditions."[8] The color can range from white to violet. A reddish-violet discharge might be likened to a fire at night and yet resemble a dark cloud at day; a brighter one might always appear as fire.

In September 1949, while in Yellowstone National Park, a Mr. William B. Sanborn witnessed a hazy patch of blue light, about 50 yards wide and five times as long, move across a marsh toward him at several feet per second:

> When the patch was but a few yards away, I noted a

---

* The investigator made no reference to Moses' experience.

> sudden calm in the air and a marked change in temperature, as well as what I believe was the odor of ozone [indicative of an electrical field]. . . . It kept low to the ground, actually "flowing around" everything that it came in contact with, coating it with a strange pulsating light. Each twig on the sagebrush was surrounded by a halo of light about two inches in diameter. It covered the automobile and my person but did not cover my skin. There was a marked tingling sensation in my scalp, and brushing my hair with the hand caused a snapping of tiny sparks. . . . I obtained no shock from touching any object on the ground or the outside of the car.[9]

An experience like Sanborn's is disconcerting even today. Imagine the stirrings in the hearts of the Israelites if they beheld such a phenomenon as this.

It seems that natural electrical activity was rampant during the time of the Israelite Exodus. If a natural electric field encompassed some such terrestrial body as the Tabernacle, there might well result the generation of a *super corona* from electric field distortions over the large assemblage of conducting material (although electrically conducting material is not absolutely necessary for the effect).

With an intermittent presence of the field, the fiery cloud would come and go above the Tabernacle at irregular intervals. When it was present, the Israelites encamped; when it was not, they moved.

Inside the structure, with all its metallic trappings, there would have been other electrical activity when the field was present. Sometimes, depending on the strength of the field, this activity -- arcing and coronas -- occurred throughout the interior. Moses was then unable to enter because "Jehovah's glory filled the Tabernacle." Because of capacitance effects, discharge phenomena would continue even after the field died out.

It is possible, too, that coronas would appear over the heads of anyone in the area of increased field strength, the body contour contributing to the effect -- viz., the spirit settling down upon the seventy visitors and the glow at other times upon Moses' face (Exod. 34:29-30).

This interpretation is perhaps supported by the report of an unusual thunderstorm in 1880:

> [A] tremendous peal of thunder shook the houses of Clarens and Tavel to their foundations. At the same instant a magnificent cherry-tree near the cemetery . . . was struck by lightning. Some people who were working in a vineyard hard by saw the electric "fluid" play about a little girl who had been gathering cherries and was already 30 paces from the tree. She was literally folded in a sheet of fire. The vine-dressers fled in terror from the spot. In the cemetery six persons, separated into three groups, none of them within 250 paces of the cherry-tree, were enveloped in a luminous cloud. They felt as if they were being struck in the face with hailstones or fine gravel, and when they touched each other sparks of electricity passed from their finger-ends. At the same time a column of fire was seen to descend in the direction of Le Chatelard, and it is averred that the electric fluid could be distinctly heard as it ran from point to point of the iron railing of a vault in the cemetery. . . . [N]either the little girl, the people in the cemetery, nor the vine-dressers appear to have been hurt. . . .[10]

Returning to the seventy elders, what else did they experience? Not all of the possible effects of such electrical phenomena upon living organisms are known. Variations in emotional responses and the ability to move around are possible.* According to A. R. Sheppard and M. Eisenbud, "Both electric and magnetic fields of relatively low strength appear to affect higher functions of the brain involving cognition or the perception of time."[11] By way of example, they report that "clinical studies of 45 employees of a Soviet 400-500 kV [electrical] substation . . . resulted in a diagnosis of neural pathology in 28 cases. 'Functional disruption of the central nervous system' was diagnosed in 26 cases . . ."[12] Of relevance to "the spirit settling upon the Seventy" is the following: "Persons in fields greater than several kilovolts per meter are subject to spark discharge. . . . There is . . . evidence that these discharges are capable of producing neurological effects, including

* See discussion in Chapter XII.

altered electroencephalograms (EEGs). . . ."[13] Perhaps an *altered state of consciousness*? In some way, because of the electric field, the Seventy "acted like prophets."

There were areas beyond the Tabernacle where the effects of the field were significant, as evidenced by the other two individuals behaving like the Seventy. But people are different, and some are doubtless more susceptible than others to such a stimulus.

According to the biblical information, Moses' contact with the Burning Bush possibly was his first exposure to electricity. At any rate, his curiosity, prompting his approach to it (Exod. 3:3), implies a lack of complete familiarity. Most of the manifestations involving the Tabernacle probably were unexpected and given an *ad hoc* interpretation. The anticipation of communication from Jehovah by way of the Ark even before it was built seems to be an exception (Exod. 25:22).

One item of wonder remains: the sounds emitted by the Ark and, earlier, the Bush.

If the two Cherubs were electrically insulated from the Ark's cover, on which they were mounted, there existed the possibility of voltage differences (both during and after field presence) between Cherubs and between each Cherub and the Ark; this would depend on the geometry and proximity of the gold overlay at their points of attachment. In addition to corona effects, arcing could occur between Cherubs and between each Cherub and the Ark.

Discharges from the Ark and Bush would have been accompanied by the now-familiar hissing, buzzing and crackling sounds that sometimes characterize such discharges. But in what state of consciousness was Moses the Prophet at those times? And what else did he hear?

We can only wonder.

Solomon's Temple

Nearly a half-millennium after the Exodus, Solomon erected in Jerusalem a more permanent edifice honoring Jehovah: the Temple. Like the Tabernacle, it was splendid in its metallic finery; gold overlay proliferated throughout the interior -- on structure and furnishings alike. Immediately upon completion, Divine approval was presumed at the probably unexpected reappearance of the Cloud of Jehovah, which filled the Temple in a repetition of what Moses had witnessed in the wilderness:

> Finally all the work that Solomon had to do for the house of Jehovah was at its completion . . . and it came about . . . the house itself was filled with a cloud, the very house of Jehovah, and the priests were not able to stand to minister because of the cloud; for the glory of Jehovah filled the house. . . . It was then that Solomon said: "Jehovah himself said he was to reside in the thick gloom" (II Chron. 5:1, 13-14, 6:1).

That the Cloud was associated with an electric field is attested by what followed its appearance in the Temple. The field was stronger than in Moses' day, and the greater strength led to a "catastrophic attempt" of nature to neutralize the voltage difference responsible for the field. Solomon offered a lengthy prayer of dedication, and then, "as soon as Solomon finished praying, the fire itself came down from the heavens and proceeded to consume the burnt offering and the sacrifices" (II Chron. 7:1). Unwittingly, the people of Israel had witnessed an electrical discharge from the sky to the earth.

A Scourge in the Wilderness

Returning to the time of Moses, the subject of sky-to-ground discharges may be the key to solving one more puzzle associated with the Camp of Israel:

> [And] the whole assembly of the sons of Israel began to murmur against Moses and Aaron. . . . [T]hey then turned toward the tent of meeting; and, look! the cloud covered it, and Jehovah's glory began to appear. . . . Then Jehovah spoke to Moses, saying: "You men, rise up from the midst of this assembly, that I may exterminate them in an instant."
>
> Moses said, ". . . the plague has started!" . . . And those dead from the scourge amounted to fourteen thousand seven hundred (Num. 16:41-2, 44-6, 49).

These deaths were much too sudden and circumscribed (Moses merely moved out of the way) to have been caused by disease. Yet what could have acted on such a scale so quickly? Could sky-to-ground discharge have anything to do with deaths that evidently were so unspectacular -- deaths not attended by "fire" or other visible signs?

There *was* one visible sign: The Cloud had made an appearance -- the electric field was back. Modern observation again provides a clue:

> Should the electric field [responsible for lightning effects] not fully develop due to moderating space charge influences in the atmosphere or ground-related aspects, then flashless discharges may occur. . . . [There is documented] a flashless discharge occurrence which caused the deaths of several people and animals. No discharge was noticed by the nearby observers. Yet evidence of high-current discharge was verified by examination of the immediate area.[14]

The errant Israelites most likely suffered a similar fate: They were electrocuted by multiple strokes of "invisible lightning."

Biblical Precedents for Uniformitarian Behavior?

Holy scriptures of antiquity often had their origins in the inexplicable. They contain records of events that could be perceived only as miraculous. Electricity, not to be comprehended for millennia, provided ample cause of both wonderment and fear. Its beautiful manifestations would come and go as a spirit without substance; against its danger there was little defense. Priestly wisdom held that succeeding generations should remain forever aware not only of the marvels, but especially of the judgments experienced by their forebears, lest in ignorance they should by disobedience become subject to similar retributions.

The pattern of thaumaturgical literature therefore was incomprehension, then transcription and "exegesis," then veneration. Something of the like has reared its head in science over the past 140 years, and the writings of uniformitarians have to be sifted as carefully as those of the ancients.

The good news is that glimmers of fact can be found in both ancient and uniformitarian documents. The reader simply has to jettison old assumptions in either case, in order to separate the fact from the fiction.

REFERENCES

1. F. Llewellyn-Jones, *The Glow Discharge* (London 1966), p. 1.

2. J. R. Perkins, "Some General Remarks on Corona Discharges," *Engineering Dielectrics* Vol. 1 (Philadelphia 1979), pp. 3-4.

3. Ibid., pp. 3, 5, 9.

4. Leonard B. Loeb, *Electrical Coronas: Their Basic Physical Mechanisms*, University of California Press (Berkeley 1965), p. VII.

5. Immanuel Velikovsky, *Worlds in Collision*, Doubleday (Garden City 1950), pp. 267-8.

6. Ibid., pp. 163-7.

7. James Dale Berry, *Ball Lightning and Bead Lightning*, Plenum Publishing (New York 1980), p. 197.

8. Loeb., op. cit., p. 10.

9. *Natural History*, 59:258-9 (June 1950).

10. *Nature*, 22:204 (July 1, 1880).

11. A. R. Sheppard and M. Eisenbud, *Biological Effects of Electric and Magnetic Fields of Extremely Low Frequency*, New York University Press (New York 1977), pp. 2-9.

12. Ibid., pp. 2-15.

13. Ibid., pp. 2-25.

14. Barry, op. cit., p. 196 (emphasis added).

Biblical quotations in this chapter are from the New World Translation.

AFTERWORD

The Uncertain Depth of History

At the dawn of the nineteenth century, Western man had no real conception of the depth of the history of man or of the earth. Facts that were available were not adequately understood. Ideas about origins and prehistory thus conformed not so much to facts as they did to authority. The authority, of course, was the Bible.

The Masoretic text (Old Testament), prepared between the Sixth and Tenth Centuries A.D., has been accepted universally as the only authentic Hebrew Bible. An earlier version is the Septuagint, a Greek translation from the original Hebrew that dates from the Third and Second Centuries B.C. There are differences in the two that produce different dates for events in the earliest part of the biblical record. (These differences also reflect a contradiction between the Masoretic text and the New Testament Book of Luke, 3:36.) Primary events that differ in ascribed dates are the Creation, with a Masoretic-derived date of 4004 B.C. vs. a Septuagint-derived date of c. 5650 B.C., and the Flood, with a Masoretic-derived date of c. 2460 B.C. vs. a Septuagint-derived date of c. 3400 B.C.

These dates, however, were soon rejected (although the Septuagint was not specifically part of the debate) as researchers began to realize that the crust of the earth spoke of a much greater age than some 6000 years. And with the entrenchment of uniformitarianism and Darwinism, not only the dates but even the events in early biblical history were set aside; specifically, major catastrophes, such as the Flood, were regarded as fictions. This attitude prevailed in spite of abundant physical evidence, recognized even by Darwin, that in fact such upheavals did occur.

Little direct challenge to accepted "scientific dates" has been mounted in this work, but an implicit challenge is lodged in every chapter even though modern dating techniques, on the surface, appear to present a formidable obstacle to many of the propositions presented herein.

The old saying, "You can't tell a book by its cover," often applies to situations as well, including this one. The bedrock of much orthodox chronology is radiometric dating. This method is used whenever minerals are available to which it can be applied. Thanks to this process, "scientifically" ascribed ancient dates are virtually sacrosanct. Radiometric dating, however, rests on assumptions that are as unfounded as those underlying Darwinism.

The uniformitarian technique of radiometric dating goes like this: One element ("parent") radioactively decays at a measured rate into another ("daughter"). By the relative amounts of the parent and daughter elements in a given mass of material, the age of the material is easily calculated vis-à-vis the "known" decay rate, assuming that this rate has always been equal to its present value. And we are led to believe that, indeed, the rate *has* remained constant since the original element or compound was formed -- thousands of years for carbon-14 and millions of years for many other materials. We are assured further of the immunity of this dating process to external interference. Finally, one of the most convenient characteristics of the entire procedure was declared to me once in the classroom, quite matter-of-factly, by the chairman of the geology department of a major university: "If the dates don't fit your theory, throw them out -- use only the ones that do."[1] As high-handedly arbitrary as it sounds, such manipulation of the evidence is not an uncommon practice: There are documented cases of deliberate deceit to preserve dates actually refuted by objective analysis.[2]

Aside from such editing, and contrary to uniformitarian declarations, there is evidence of radioactive decay rates being altered by virtue of changes in the chemical compounds of which the subject elements are a part. Similar inconsistencies can result from variations in pressure.[3] Articles have appeared in *Nature* and in The *American Journal of Physics* claiming that little or no justification exists for assuming the stability of decay rates over geological time. If these articles state a just case, the possibility of accurate radiometric age calculations vanishes like a Will-o'-the-wisp, and accepted geological dates must be viewed as highly suspect.

Some interesting anomalies rivet the attention. Shells of living mollusks have been dated by carbon-14 at 2,300 years. Moon rocks have been dated by different radiometric methods at two to 28,000 million years of age -- the latter being several times the presumed age of the solar system. (With an insight into the nature of ancient lunar disturbances, Velikovsky actually foresaw this anomaly *prior* to the lunar landings.[4]) All around the world, lava flows that have occurred during the last two centuries have been potassium-argon-dated at hundreds of millions of years.[5]

How old is Earth? We don't know. It may be billions of years old, but radiometric dating is no proof of this.

In the absence of appropriate radioactive materials, geological dating becomes even more precarious. According to Derek Ager, geological phenomena "must always be measured (in the absence of anything better) against the scale provided by organic evolution"[6] -- that is to say, against a Darwinist geological calendar. A more imprecise, inaccurate scale would be difficult to conceive.

During the last hundred years, we have moved 180 degrees from the nineteenth century perception of time. Far from being too shallow, our perception today of the earth's history could better be compared to the Bottomless Pit. To suggest that dinosaurs lived significantly less than 65 million years ago is to be relegated to the fringe.* To suggest that they lived as recently as 65,000 years ago would be seen by many as reason for assignment to the lunatic fringe. Yet the latter figure is not out of the question. The uncertainty of underlying assumptions and the unreliability of so many chronological observations[7] are cause to believe that the position on time taken by orthodox science is once again based more on authority -- that of a favored philosophy -- than on facts.

---

* Such a position, of course, is merely indicative of the relative size of support for a given viewpoint. Darwinism also was once far removed from "orthodox science."

What dates, then, or magnitudes of elapsed time, are realistic? Unfortunately, this is a matter of great uncertainty, other than to say that the geological eras were far more recent than historical geology claims and far older than creationists allow. Indeed, some of the actual "eras" overlap recorded history and at the same time predate creationist perceptions of beginnings. I have argued, in fact, that there was intelligent life on Earth prior to the mythical "Beginning."

Realistic absolute dates are hard to come by at present. There has been limited speculation based on catastrophist assumptions, most notably by Velikovsky, who commented: "I am not in a position to point to the century or even the millennium when the Universal Deluge took place, but it must have happened between five and ten thousand years ago, probably closer to the second figure."[8]

Velikovsky's upper limit is compatible with the conventional estimate of the original date of the founding of the world's oldest known city, Jericho, believed to have been first established circa 9000-8000 B.C.[9] If the city has any relevance to the dating of the Flood, it is the fact that it could help establish the latest possible date that the waters receded.

Concerning the more remote past, Roger Ashton expresses an opinion on the time of the appearance of the first planet-god, Saturn, i.e., "The Beginning": "[C]rucial characters of gods, or the primordial god, did in fact originate at least 20,000 years ago."[10] Again, the line drawn on the scale of time is but a wavering one.

All dates, then, are highly speculative, and there are few enough even of these. There remains staggeringly much to be uncovered -- and against the hostile resistance of the uniformitarians.

For now, our study of origins must be concluded with a display of the few blocks that we have hewn from the record in the rocks and from ancient reports, both oral and written. Our results, based on explanations embedded in catastrophic processes (celestial, geological, biological, cultural), are summarized and depicted in the accompanying charts, "Catastrophist Geo-historical Column" and "Summary of Cataclysms."

Our planet has been jolted and pummeled. Our environment has been molded and remolded. Species have been repeatedly created and destroyed. Celestial neighbors have come and gone.

One violation has engendered another. Regimes have been overturned, starting with extraterrestrial disturbance: the atmosphere,

Earth's surface, the crust, all life. New species arose as a result; in some instances, these bore little resemblance to their progenitors. Moreover, we find written in the rocks not just the origin of species, but the sudden birth of entire genera, containing many species.

There is no evidence in Earth's crust that the processes of gradualism are the ruling factors in the origin of species; gaps between fossil types abound, as they should not do if gradualism had any answers. Neither is there any uniformitarian evidence in the biosphere. Illusions of evidence are created by empty extrapolation. Gradualism measures an inch and stretches it to light years.

All kinds of records, natural and historical, bespeak the repeatedly collapsing schemes of nature and culture, followed each time by new life, new interrelationships, new environments, new institutions. Man himself has been biologically and psychologically altered by the more recent of the disturbances.

We must break the shackles of uniformitarianism because the history of our planet is one of repeated violent metamorphosis, of repeated revolution in the biosphere, of repeated revolution in the lithosphere. It is a history of cataclysm.

Further pursuit of interdisciplinary studies within a catastrophist framework holds out to the newcomer the certainty of unexpected new vistas and perspectives. A continued survey of existing catastrophist hypotheses would take us far afield from the issues of this book.

For those of us who have pursued catastrophist logic and research for decades, the road ahead is even more exciting: What we know about ourselves and our planet is negligible compared to what remains to be discovered. We can anticipate that the coming years will be replete with new insights into the pasts of mankind, Earth, our solar system, and the rest of the universe. Space probes continue to produce suggestive new evidence, if only it can be sifted and weighed in the right way. So do Earth probes.

What cannot be stressed enough is that even the most seemingly incredible testimonies bequeathed to us by our predecessors in antiquity suggest yet another kind of probe that we should be doing. Ancient man's words are worthy of more scrutiny by far than the uniformitarians are willing to grant. Although early man's understanding was limited, his intellectual capacity was no less than our own. Some, even, of his most fantastic memoirs and chronicles are good-faith reports of observed phenomena, and judgments of their integrity cannot be based on their conformance to modern theories or opinions.

In truth, the "myth" of our ancestors is sometimes more rational than the "science" of our contemporaries, much of which will be tomorrow's myth.

Time and questing minds will show that the record of cataclysm is historical. It is the only significant record in the rocks. It is the record all of us should be trying to read.

CATASTROPHIST GEO-HISTORICAL COLUMN

(Tentative - No Time Scale Implied)

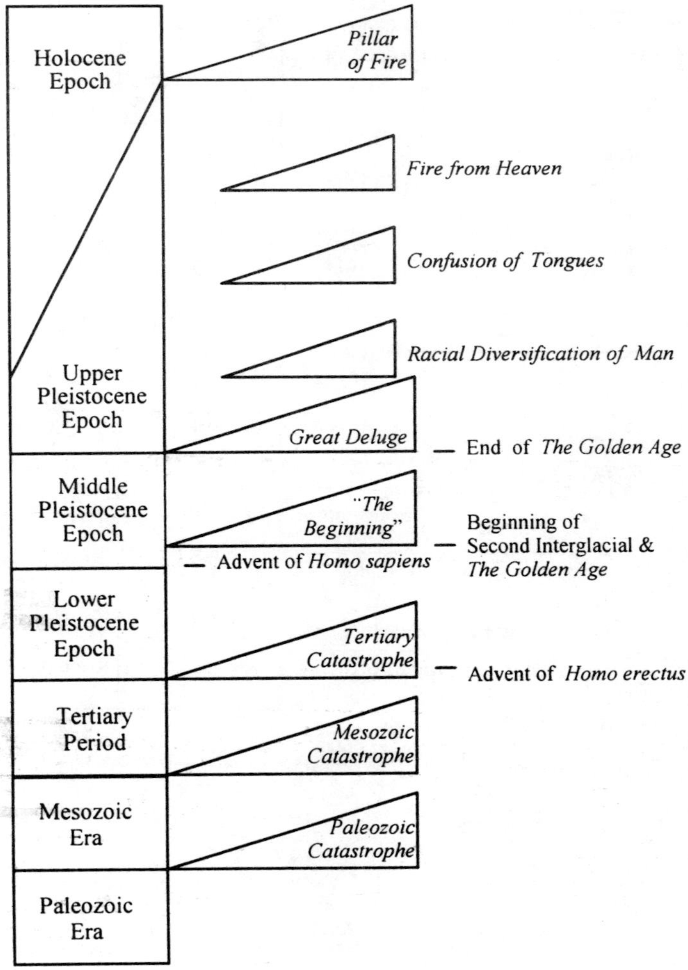

SUMMARY OF CATACLYSMS

(TENTATIVE)

CATACLYSM	TIME OF OCCURRENCE	ASSOCIATED EVENTS	EXTENT	REF.
Paleozoic	End of Paleozoic Era	Beginning of Geological Record	Global	Chapter V
Mesozoic	End of Mesozoic Era	Extension of Geological Record	Global	Chapter V
Tertiary	End of Tertiary Period	Extension of Geo. Record *H. erectus*	Global Old World	Chapter V Chapter VII
The Awakening	End of Lower Pleistocene	Magnetic Reversal *H. sapiens*	Global Old World	Chapter VII Chapter VII
The Beginning	Beginning of Second Interglacial	Saturn Nova 'Let there be light.'	Global or Near-Global	Chapter X
Great Flood	End of Middle Pleistocene	Saturn Nova Races of Man	Global Old World	Note 1 Chapter XI
Confusion of Tongues	Between 10th & 3rd Millennium B.C.	Tower of Babel	Global or Near-Global	Chapter XII
Fire from Heaven	c. Beginning of 2nd Millennium B.C.	Sodom and Gomorrah	Middle East, So. America, Other	Chapter XIII
Pillar of Fire	Middle of 2nd Millennium B.C.	Israelite Exodus	Global	Note 2
Day of the Lord	7th-6th Centuries B.C.		Global	Note 3

NOTES:

1. See Immanuel Velikovsky, "On Saturn and the Flood," *KRONOS* V-1 (Fall 1979).

2. See Immanuel Velikovsky, *Worlds in Collision*, Part I, (Garden City 1950).

3. Ibid., Part II.

REFERENCES

1. Fall 1983.

2. See Immanuel Velikovsky, "Ash -- A Historical Record" for a remarkable example of this practice: *Pensée* IV-1 (Winter 1973-1974), pp. 5-19.

3. Henry Faul, *Nuclear Geology* (New York 1954), p. 10.

4. See Thomas Ferté, "A Record of Success," *Pensée* II-2 (May 1972), p. 13.

5a. A. F. Kovarik, "Calculating the Age of Minerals from Radioactivity Data and Principles," *Bulletin No. 80*, National Research Council (June 1931).

b. M. Keith and G. Anderson, "Radiocarbon Dating: Fictitious Results with Mollusk Shells," *Science* 141 (1963).

c. Funkhouser, Barnes and Houghton, "The Problem of Dating Volcanic Rocks by the Potassium Argon Method," *Bulletin of Volcanologique* 29 (1966).

6. Derek V. Ager, *The Nature of the Stratigraphical Record*, 2nd ed. (London 1981), p. 71.

7. See, for example, Velikovsky, "The Pitfalls of Radiocarbon Dating," *Pensée* III-2 (Spring-Summer 1973), pp. 12-14, 50.

8. Ibid., p.13.

9. Jacquetta Hawkes, *Atlas of Ancient Archaeology*, McGraw-Hill (New York 1974), p. 198.

10. Roger Ashton, "The Genie of the Pivot," *KRONOS* X-l (Fall 1984), p. 21.

APPENDIX

Archaeo-Electrics in the Origins of Christianity

Like the universe of which he is a part, man is an electrical creature. The natural currents within our bodies are small, but their role is extremely significant: They govern our very lives.

These minute currents can be disrupted or overridden. In particular, "[E]xtremely low frequency (ELF) fields have the capacity to penetrate buildings and living tissue and hence are potential biological stimuli."[1] When this happens, our perceptions can falter, and our reaction and response times might prove to be beyond our control. Such stimuli might have many unpredictable effects upon a person, but there are a few effects that can be described from experience.

As to perception, certain fields can excite the human auditory system in such a way that the brain perceives "sounds" that do not exist:

> Certain individuals are extremely sensitive to the sounds and as yet undefined effects of electromagnetic fields to which they are exposed. . . . Nurses who work in mental institutions describe patients who are always complaining and trying to get away from the "terrible noise." Cotton in the ears did no good, but certain rooms or areas were more quiet for them [an electrical field null point?].[2]

Such "sound" would be omnidirectional; when "heard," it would seem to come from everywhere. It apparently seems to fill any closed structure in which the hearer finds himself.

In a more passive case, we have seen that humans are susceptible instruments for the phenomenon of corona discharge -- the luminous electrical effect appearing in proximity to a physical surface within an electric field.[3]

Persons in fields greater than several kilovolts per meter are subject to spark discharges. . . . A person distorts an electric field nearby so that the actual field at the surface of the head, arm or knee may be stronger than the original field. . . . [For] a prolate spheroid [approximating the human body] the field at the top [head] is increased by a factor of 15.[4]

Thus, given the geometry of the human torso, immersion within some specific electromagnetic field might therefore even emblazon a person's countenance, or perhaps give one the appearance of wearing a hood of fire.

Always accompanying such a discharge is a buzzing, crackling or hissing sound. This "sound generated by . . . coronas can be heard by ear, and for power lines operating in the 500 Kv to 1200 Kv region that are situated near highly populated areas, the audible noise can be so severe that a change in the transmission line route is demanded."[5]

One can imagine the possible auditory effect of a corona just above or surrounding the ear. Like the field-induced pseudo-sound which might accompany it, it would seem to be everywhere.

There is yet another noteworthy effect observed in a previous chapter,* to wit: Electrical stimulation of the motor strip can evoke vocalization, but such vocalizations do not constitute real words. This kind of vocalization has been described as *inappropriate or garbled speech*. Moreover, it is not necessary that the stimulation be a direct electrical application, inasmuch as ELF electromagnetic fields can have similar effects.[6,7,8,9]

Presumably, such "speech" might sound to an untrained ear like another language -- another *tongue*.

To experience simultaneously all of these effects (noise, fire, "tongues") could prove to be overwhelming -- or might it instead become an emotional high?

---

* See Chapter XII.

These observations make a provocative background for a rereading of the biblical account of the coming of the Holy Spirit on the Day of Pentecost:

> And when the day of Pentecost was fully come, they were all with one accord in one place. And suddenly there came a sound from heaven as of a rushing mighty wind, and it filled all the house where they were sitting. And there appeared unto them cloven tongues like as of fire, and it sat upon each of them. And they were all filled with the Holy Ghost, and began to speak other tongues, as the Spirit gave them utterance (Acts 2:1-4).

Whatever the source might have been for the presumed electrical activity underlying this passage, two millennia ago it was not understood that these various effects were manifestations of a single phenomenon. This fact could have been recognized only from experience.

REFERENCES

1. M.A. Persinger, H.W. Ludwig, K.P. Ossenkopp, "Psychophysiological Effects of Extremely Low Frequency Electromagnetic Fields: A Review," *Developmental and Motor Skills*, 36 (1973), 1131.

2. J.B. Beal, "Electrostatic Fields, Electromagnetic FIelds, and Ions - Mind/ Body/Environment Relationships," *Biologic and Clinical Effects of Low Frequency Magnetic and Electric Fields*. (Springfield 1974), 15-16.

3. J.R. Perkins, "Some General Remarks on Corona Discharges," *Engineering Dielectrics*, 1 (1979), 5, American Society for Testing and Materials.

4. A.R. Sheppard, M. Eisenbud, *Biological Effects of Electric and Magnetic Fields of Extremely Low Frequency*, (New York 1977), 4-25.

5. R.T. Harrold, "Acoustical Techniques for Detecting and Locating Electrical Discharges," *Engineering Dielectrics*, 1 (1979), 345.

6. H. Whitaker and H.A. Whitaker, *Studies in Neurolinguistics*, Vol. 1, (New York 1976), pp. 120-121.

7. J. Van Buren, C.L. Li, G. Ojemann, "The Fronto-Striatal Arrest Response in Man," Electroencephalography and Clinical Neurophysiology, Vol. 21, 128.

8. M.A. Persinger, *ELF and VLF Electromagnetic Field Effects*, (New York 1974), 4.

9. Ibid., p.276.

INDEX

—A—

—B—

—C—

—D—

—E—

—F—

—G—

—H—

—I—

—J—

—K—

—L—

—M—

—N—

—O—

—P—

—Q—

—R—

—S—

—T—

—U—

—V—

—W—

—Y—

—Z—

ABOUT THE AUTHOR

James E. Strickling holds degrees in electrical engineering and marketing. He is also a *cum laude* graduate in interdisciplinary studies (natural sciences, ancient history, philosophy). He is a former Member of the Technical Staff at one of AT&T's manufacturing facilities (now retired), and he teaches mathematics at a Georgia university.

Mr. Strickling's publications have appeared in the interdisciplinary journals *KRONOS*, *Aeon, Catastrophism and Ancient History*, and *The Society for Interdisciplinary Studies Workshop* in Great Britain.

A former creationist, Mr. Strickling now numbers himself among the "secular catastrophists," scientists and scholars who argue that the earth has been shaken and shaped in the past by global disturbances. In this work, he presents an unprecedented critique of the creation-evolution controversy from the perspective of an unbiased observer outside both of the traditional camps. He argues that an entirely new worldview is necessary to adequately understand the history of our planet and its inhabitants.

Strickling also deals at length with natural electrical phenomena occurring in antiquity and recorded in ancient documents, the Bible among them. Such phenomena would not be obvious to the typical anthropologist or historian, but thanks to his electrical engineering background and an unfettered perspective, Strickling has shed a surprising new light on a number of ancient, and previously misunderstood, texts.